gardening for everyone

gardening for everyone

GROWING VEGETABLES, HERBS, AND MORE AT HOME

Julia Watkins

ROBINSON

First published in the US in 2022 by Mariner Books,
an imprint of HarperCollins Publishers

First published in Great Britain in 2022 by Robinson

1 3 5 7 9 10 8 6 4 2

Book design by Ashley Lima
Illustrations on pages 65, 77, 79, 125, 126, 129, 131, 134, 183 by Allison Meierding; pages 237, 245, 250, 260, 263, 271, 280 artnLera/Shutterstock; page 272 Bodor Tivadar/Shutterstock; pages 9, 73, 89, 107, 157, 175, 231, 265, 282 Elena Pimonova/Shutterstock; page 274 Liliya Shlapak/Shutterstock; pages 240, 241, 243, 254, 259, 276 mamita/Shutterstock; pages 235, 238, 247, 248, 251, 258, 266, 267, 268, 281 MicroOne/Shutterstock; pages 244, 253, 257, 262, 277 NataLima/Shutterstock; pages 234, 270, 279 Natalya Levish/Shutterstock; page 246 Olga Lobareva/Shutterstock; page 273 Varlamova Lydmila/Shutterstock

A CIP catalogue record for this book
is available from the British Library.

ISBN: 978-1-47214-692-2

Printed and bound in Italy by L.E.G.O. S.p.A.

Papers used by Robinson are from well-managed forests
and other responsible sources.

Robinson
An imprint of
Little, Brown Book Group
Carmelite House
50 Victoria Embankment
London EC4Y 0DZ

An Hachette UK Company
www.hachette.co.uk

www.littlebrown.co.uk

How To Books are published by Robinson, an imprint of Little, Brown Book Group. We welcome proposals from authors who have first-hand experience of their subjects. Please set out the aims of your book, its target market and its suggested contents in an email to howto@littlebrown.co.uk.

HUNTER
HUNTER

contents

PLAYING

PLANT PROFILES

introduction

I grew up in a small town in South Carolina, a stone's throw from my grandparents' house, where as a kid I spent a lot of time. This was in the 1980s but, whenever I was there, it felt like something out of the 1950s—or even the 1800s. My grandparents lived a slow, simple life, with a calm, steady routine. They cooked every meal from scratch—no store-bought sauces or just-add-water mixes for them. Most ingredients were picked fresh from the big, sumptuous garden just outside their back door. And almost nothing was ever left to waste.

My grandfather dug a vegetable plot nearly the size of his back garden almost every year of his life. He'd then go and dig his mother's garden, which he also did nearly every year of his life. She made it to age 102 and was still tending her garden the spring she passed away. My grandfather's gardens were immaculate. Each crop occupied two long rows, with seedlings spaced perfectly apart. Tomatoes were staked in a weave string trellis, and corn always stood tall and proud. Watermelons and cantaloupes slumped to the same side, while lines of cabbage, spinach, and other greens neatly framed the cucumbers, courgettes, radishes, turnips, beetroot, potatoes, green beans, black-eyed beans, summer squash, and okra.

His gardens were kitchen gardens—in every sense of the term. Close to the kitchen, tended daily, and grown for home, with he and my grandmother making frequent trips out back at mealtimes and whenever someone stopped by for a visit. I always loved just hanging out in the garden, walking idly among the rows. My grandfather did too, and he even placed a rickety old swing by a nearby tree, with room for three, if you didn't mind the squeeze. Whether you were harvesting a melon or passing the time, my grandfather's garden was the place to be.

Life revolved around the garden. By the back door, my grandparents kept a broad harvest table, often lined with tomatoes, melons, and sacks of black-eyed beans. If we weren't growing food, we were eating it. Whenever my mom dropped by, she'd sit at the table with my grandmother, playing catch-up while snapping green beans and snacking on sliced, salted tomatoes. I can't remember a weekend where we didn't have collard greens, creamed corn, and potatoes for lunch with the whole family!

My taste buds can attest—there's nothing more scrumptious than food picked fresh. As a kid, I'd clamor for my grandparents' veggies over ice cream and cookies. Truly, I would.

When I'd grown up and left home, I brought gardening with me. It wasn't planned so much as it just happened. Wherever I went, I nearly always found myself near a garden. Living in Charleston as a college student, I took a job as a clerk in a gardening shop on King Street. Working at a ranch in Colorado in my early twenties, I ended up teaching gardening to children spending the summer there. Spending time after college in Asheville, where my great-grandmother had a tiny old cabin up in the mountains, I found work growing herbs at a nearby greenhouse.

Then I joined the Peace Corps and was packed off to West Africa, where I lived in a remote village in Guinea for two years. I'd barely settled in before I found myself helping farmers plant trees, graft mangoes, build mud stoves, and keep bees. There, too, I kept a garden, starting with seeds my grandparents sent in the mail. (I later learned that shipping seeds internationally was a big no-no—but at the time I couldn't have been more excited.) I also sourced seeds from markets in Conakry, Guinea's capital, some of which came from elsewhere in West Africa—Senegal, Mali, Ivory Coast, and Nigeria.

I built raised beds, staked out a compost heap, and planted a "live" fence made from moringa tree saplings planted and woven together. A few of the villagers turned me on to it, as a more sustainable alternative to split rails and barbed wire, which I wasn't likely to find much of in any event. A live fence is a useful innovation in a place like Guinea—where deforestation is all too common and what wood there is gets used for cooking.

As the garden grew, I made sure to share fresh-picked ingredients for salads and stews with everyone who visited.

After the Peace Corps, I came home and moved to Washington, D.C., where I got married and lived with my husband in one of the relatively few farmhouse-style detached homes in the northwest part of the city. Like many lots in D.C., ours was tiny, and we experimented with a little container garden in the mostly bricked-over back garden. We grew tomatoes and peppers in a raised bed and a smattering of herbs in terra-cotta pots.

Soon after our son was born, this little kitchen garden was more than enough to fulfill my itch to grow food, my hands full with changing diapers and nursing a baby. But what a delight, the following summer, watching our little guy take some of his very first steps over to the tomatoes, poke and pull at the plants, and tip over his little toy watering can, sometimes even getting the water on the tomatoes and not himself—or me!

We soon moved further out from the city and, with a little more land, built a few raised beds in our new garden. My daughter was born not long thereafter and, as both kids grew, they each got a little square bed to try their own tiny hands at gardening. They chose vegetables they liked to eat. Then they sowed just enough seeds to grow them in the space they had. It was a fun way to loop them into gardening, even if bunnies and skunks ate most of our food!

We gardened for years in this small but meaningful way, teaching our children the joy of growing food while sharing our enthusiasm for nature, worms, bees, and soil. We moved again, first out to California for a few years and then finally into our forever home just north of Chicago. And it was here we decided to go big, planting nine raised beds, a handful of pots, two pollinator gardens, and as many wildflowers as we had space to grow.

This garden is, to me, a true kitchen garden—just ten feet from the back door that opens into our kitchen. While cooking, I can peer out the window and see the little dream we created growing in the garden. I used to think there was nothing better than a view of the ocean. I still think that way, unless, of course, I can peep out the window at a garden. It gives me joy. It gives me peace. And I'm not sure there's anything better for the soul than to spend time out back, my hands as busy as can be, listening to the birds chirp, getting covered in dirt, and feeling connected to all the best parts of my life. I tend my garden and my garden tends me.

Of course, I garden for a lot of very practical reasons, too. Drawing at least a portion of our food from our garden is resourceful—a step toward self-sufficiency. It saves us money, supports all kinds of beneficial insects and birds, and helps keep pesticides and the dirt and grime of food transportation out of what we eat.

But what I love most about gardening is the variety of it. No two gardens look alike. For what starts with some basic, timeworn principles quickly gives way to the artistic flair and aesthetic sensibilities of the gardener. My garden might never be *exactly* the way I want it, if only because I seem to want something new whenever I'm in it. But my garden is very much me. It reflects my values, my personality, my lifestyle, my circumstances. And your garden, however you choose to do it, should be very much you.

If money's tight or sustainability's a goal (or both), you can upcycle natural and secondhand materials for garden beds, collect rainwater in barrels for watering, save seeds and trade with friends, repurpose scrap wood for a compost bin, grow crops from seed, and share tools and even water with your neighbors.

If health is your thing, you can commit to organic gardening, including making your own compost and fertilizer from organic materials and managing pests with beneficial insects and companion planting. If you're short on space, there's nothing keeping you from trying out a few window boxes, or stationing pots on the patio or out on the fire escape. A raised bed or two could be tucked into even postage-stamp-sized gardens. If you're really cramped, there's always vertical gardens. And you can even grow inside by a window—I'll show you how!

This little book will walk you through the basics of planning, building, planting, tending, and harvesting an organic kitchen garden—with helpful tips for doing so sustainably, on a budget, and in spaces big and small. For the most part, it sticks with the most common plants, showing you how to sow, grow, and harvest them, but it doesn't stop there, as I think every gardener should know a thing or two about native plants and edible flowers, many of which make harmonious companion plants for other vegetables and herbs.

In writing this book, I've tried to draw on all I've learned from the wonderful gardens and gardeners I've known in my life, from my grandparents on through the farmers in Guinea to my own son and daughter, carefully culling and curating all that experience into something that could guide the most ardent of beginners while still being of interest to the greenest of fingers.

I'm delighted to share it with you and hope you enjoy it!

how to use this book

This book is meant to be a practitioner's guide for gardeners—meaning it's less a book you read cover to cover and more a ready reference to consult before, during, and after each stage of working on your garden—that is, planning, building, planting, tending, and harvesting. That's why it's organized by stage and chock-full of easy-to-digest tips and practical DIYs. I imagine it being the kind of book you pull off the shelf when you have a question and need an answer. I also envision it being smudged with soil and stained with flowers. It's designed to spend as much time in your garden as you do.

This book also includes a section called "Plant Profiles" to reference when you feel you need specific information about a particular crop. Perhaps you'd like to know whether a plant can grow in pots or whether it has specific preferences for sun, soil, or water. Maybe you're curious whether it's better to start it from seed or to buy it as a transplant, or you're not sure when and how to harvest it. The plant profiles squeeze in as much useful information as possible to help you get to know the wants, needs, and quirks of the most common vegetables and herbs.

A resource section at the end of the book tells you where to buy common items like seeds and garden supplies, as well as more nuanced things like red wiggler worms for vermicomposting.

And, my personal favorite: This book includes a special section on playing in your garden—to help you explore and create, or to connect your children to your love of gardening—packed with fun, hands-on, creative activities and recipes. A garden is a lot of work, but that doesn't mean you—and especially your little ones—can't play in it, too.

Finally—annotate! Don't hesitate to add your own notes and ideas to the pages of this book. There's an art to marginalia, and, done right, it gives each reader a chance to make their copy of this book unique, adding to the wisdom carefully curated and compiled here to reflect your own experiences, traditions, experiments, and adventures.

PLANNING

For any project, there's really nothing I like more than diving in headfirst. But a good garden starts in your head and thrives with the right planning. This section walks you through the basics of planning to help you prepare for building, planting, tending, and harvesting a healthy, productive kitchen garden.

assessing your space and choosing a location

The first step to planning a kitchen garden is to observe the space you have in mind, checking the light and otherwise evaluating its strengths and weaknesses. Almost any place can work, but some spots are certainly better than others. Gardening outdoors is obviously preferable, but even if you're in an apartment without a garden, you have options. There are all sorts of ways to work around small spaces and partial shade, which I'll discuss a little bit here and more in the next section, "Growing in Small Spaces." For now, consider the following factors when assessing your space and choosing a location.

CLIMATE

Where you live determines your climate, and your climate determines what you can grow. If you live in Miami, you'll have a hard time growing cool-season vegetables like broccoli, cauliflower, and kale. If you live in Minneapolis, you might struggle with some warm-season vegetables like aubergines and watermelons. Understanding your climate and matching food crops suited to its conditions will save you a lot of time and effort.

In the United States, you can assess your climate by referring to your area's hardiness zone. These zones are set out in the United States Department of Agriculture's (USDA) Plant Hardiness Zone Map, which divides the country into eleven zones based on their average annual minimum winter temperature. Zones go from coldest (Zone 1) to warmest (Zone 11), with each zone being 10°F warmer (or colder) in an average winter than the adjacent zone. Hardiness zones help gardeners figure out which plants may do well in their climate. Here in Chicago, I'm in Zone 5, which is considered a cool climate by growing standards.

Although hardiness zones are helpful to vegetable gardeners, they also have their limitations. For one, hardiness to cold weather is just one factor that affects a plant's health. You also have to take into account summer temperatures, the length of your growing season, the amount and timing of rainfall, and humidity levels. Your site's microclimates, which are nuances specific to your garden, are another important factor to consider. Even in the smallest garden, variations in conditions can occur from one corner of your plot to another. An area protected by a stand of trees, houses, or other buildings, for example, will have a very different microclimate than an open area facing south. Gardeners can create artificial microclimates by installing greenhouses, cold frames, and cloches to create warmth in their gardens or to extend the growing season into the cooler months of the year.

SUNLIGHT

Most garden vegetables need a lot of direct sunlight—at least six to eight hours per day. Finding the best location requires figuring out just how much direct sunlight a given area gets. Check on your desired garden spot at one-hour intervals throughout the day, and keep a record of how much time the area spends exposed to direct sunlight, dappled light, or shade.

Keep in mind that both the duration and the angle of sunlight vary throughout the year, with the sun directly overhead at summer solstice and receding almost to the horizon by winter solstice. Track the movement of the sun and consider how to orient your garden to maximize exposure during the growing season.

The best place to site your vegetable plot will usually be the most open part of your garden, to minimize any obstacles blocking the light. Check which side of trees, shrubs, fences, and structures generally stays in the sun and which side falls into the shade as the day goes by. In the Northern Hemisphere, the southern sides will get the most sun. The opposite is true in the Southern Hemisphere. Also take note: If you're assessing sunlight in the winter, areas near or under deciduous trees will get a lot of light, but come spring and summer, much of that light will be blocked by foliage.

Sometimes, there's just no avoiding obstacles, such as if you're in a small space, or your neighbor's house or that great old oak tree tower over your yard. Measure carefully the sunlight you do have and then plan accordingly, seeking crops that manage in low light or partially shady conditions if necessary. Tomatoes, peppers, and other fruiting vegetables need lots of sun and are probably out of the question. But root vegetables such as carrots, radishes, and beetroot can make do with four to six hours of sunlight per day, while leafy greens and herbs like lettuce, kale, Swiss chard, spinach, chives, coriander, and parsley can grow in partly shady conditions—that is, less than four hours of sunlight per day.

If you're short on sunlight, get creative. If an interloping tree casts too much shadow, consider trimming its lower branches. If you're planting up against a southern-facing wall, consider painting it white to reflect light. And if you're planting in pots and other containers, consider moving them around throughout the course of the day to take full advantage of the light.

WATER

After sunlight, water is the second most important element to consider. Be sure to choose a location as close to a water source as possible. A working tap and a good hose are almost always the best and most convenient options. If that's out of reach or if you're indoors, a trusty watering can that can be filled at the kitchen sink will do. You can also install a rain barrel to conserve water; just be sure you can attach a hose to it and run it to the garden.

Avoid planting in areas where water tends to pool and puddle, such as beneath downspouts and at the bottom of hills. Nothing kills plants faster than drowning their roots in water. Not only do low-lying areas not drain well, but they are also colder and more likely to create a frost-prone microclimate.

CONVENIENCE

A kitchen garden doesn't have to be right outside your kitchen door, but you'll enjoy it more if it feels like an extension of your home. I can't tell you how many times I've been in the middle of prepping a meal when I realized I needed just a pinch of fresh thyme or basil. Being able to run out back while the pot is still on the stove and snip a little bit of this and a wee bit of that is precisely what makes a kitchen garden special.

During the summer months, my garden is as indispensable as the refrigerator and the spice cabinet. I tend to leave the back door open because I'm so often scurrying from kitchen to garden and back again. Looking out the window and seeing my kitchen garden keeps it at the front of my mind and the top of my list—a ready reminder to pull any crabgrass and chickweed coming in, check the cabbage and the tomatoes for pests, and stake some slumping plants in desperate need of more support.

Of course, this is a lot to square, and these factors are not always going to line up. If you have to make tradeoffs, prioritize sun, then water, and finally convenience.

SIZE

Assuming you have a choice, make sure the area you choose is big enough to accommodate the garden you have in mind. At the same time, try not to bite off more than you can chew. Start out small, with a handful of plants you like to eat that are also suitable for growing in your climate. As you gain skills and confidence, you can always add more beds, containers, and plants. Ideally, your space will have room to expand so that as you grow as a gardener, your garden can grow with you.

Your garden's size should also reflect how many resources—in terms of time and money—you're willing and able to put into it. The bigger the garden, the more of both you'll need. Expect to spend a significant amount of time at the beginning and end of each season, preparing your garden and planting your crops in the spring and putting your garden to bed in the autumn. And then remember that gardening will be a part of your daily routine. Some days will be busier than others, but it'll never just be a weekend thing or something to turn to every now and again.

As for money, ambitious gardens don't come cheap. Building beds and buying pots, compost, mulch, and seedlings quickly adds up, so scaling a garden's size to your budget is always important. There are plenty of ways to save, of course—particularly if you can make your own compost, build your own beds, and sow plants from seed. If you're resourceful, a garden is a great place to put that skill to use!

ACCESS

Sometimes in the excitement of finding a space, gardeners forget about the importance of ensuring they have access from the outside world to their garden. Since you'll be schlepping all sorts of materials to your site, be sure to consider how you'll get them there. Do you have a gate between the driveway and the back garden? Is it wide enough to fit a wheelbarrow full of compost, mulch, or gravel? If you don't have a gate, do you have a garage door that leads to the garden? If push comes to shove, are you willing to haul everything through your house? If not, it might be worth evaluating whether there's another (better) location for your garden.

AESTHETICS

Assuming your vegetable plot will be close to the house or integrated into your garden, the way it looks will likely matter to you. If it doesn't, that's okay! For those who garden for emotional and spiritual reasons as much as for practical ones, the way your garden looks and makes you feel is important. Look for ways to ensure your garden is consistent with the style of your home and enhances the surrounding landscape. When choosing a location for your plot, look around your garden and try to envision where it will best fit in with existing structures, trees, bushes, fences, patios, and pathways.

POLLUTION

If you're considering gardening in an urban area or on a site that might be contaminated, you will need to address any soil issues to ensure a safe and productive garden. That said, the possibility of poor or polluted soil at a site should not keep you from planting a garden there. There are all sorts of things you can do to improve the quality of your soil and decrease the likelihood of exposure to contaminants. You can avoid planting in the ground altogether by growing in containers or raised beds. If planting in the ground is your preferred option, you can either use soil amendments to improve soil quality and stabilize contaminants or remove contaminated soil and replace it with certified clean soil.

HOW I DID IT

For my kitchen garden, I started by walking around our house outside a few times a day, feeling the sunlight, touching the earth, envisioning what it all could be. In one corner of our garden, there was a patch of land running along the southern side of our neighbor's fence, mostly overgrown with scrub and a few large shrubs. It would take some work, clearing all that away. But it was right outside the back door, with a tap just around the corner, bright sun overhead, and no puddles or patches of mud in sight.

The first thing we did was remove the plants growing there, leaving just the rough earth. There was room for eight raised beds, plus a ragtag fleet of pots and other containers. I've since expanded to other spots around the yard, adding butterfly gardens, a row of sunflowers, and a bed of wildflowers—with plans for more once the kids outgrow the trampoline! I installed a simple picket fence around the cleared area, laying mulch and later pea gravel paths between the raised beds.

The fence was waist-high and made of plain wood. I thought it matched the farmhouse style of our house and looked nice next to the old brick patio coming off the back of the house. I put in two gates, one directly across from the back door and the other on the far side—perfectly aligned and giving me a clear path from kitchen to garden and back again. I can't say I knew it then, but those beeline runs have become one of my greatest pleasures.

growing in small spaces

If you don't have space for an expansive kitchen garden, rest assured there's still a lot you can do to grow food. The key to growing on balconies, courtyards, patios, and small gardens is making the most of the area you have. They say limits breed creativity, and I can confirm that some of the most innovative and abundant kitchen gardens I've seen were grown on rooftops, in alleyways, on front porches, and even up walls. The key to success is in careful planning and employing creative and intensive gardening practices that allow you to make the best, most efficient use of your space. Here are a few tips that might help.

GROW IN CONTAINERS.

If growing in the ground isn't an option due to limited space, poor soil, pollution, or lack of sun, containers are a wonderful alternative. Containers allow you to customize soil for each plant's preferences. You can also move them around to chase the sun, provide shade, or create different design configurations.

There are all sorts of plants that grow well in containers, too, especially ones with shallow roots like lettuces, radishes, and spinach. Herbs are a great choice too; in fact, there are some, like mint, that spread so readily I wouldn't plant them any other way. Hanging baskets are a fun way to grow strawberries, and potatoes can be grown easily in hessian sacks or grow bags.

For containers, you can use just about anything so long as it holds soil and has drainage. If space is really limited, consider building or buying square- or rectangular-shaped containers, as they waste less space than round ones.

BEST CROPS TO GROW IN POTS

beans	peppers
carrots	potatoes
chillis	radishes
cucumbers	spinach
herbs	strawberries
lettuce	swiss chard
peas	tomatoes

GROW VERTICALLY.

There are a lot of advantages to growing vertically. Vertical gardens tend to have better air circulation and fewer pests and diseases as a result. There's also a lot less weeding and other maintenance.

Vertical gardening is perfect for growing vining type plants such as pole beans, cucumbers, and squash—just be sure to give them something to hang on to, such as a fence or trellis, as they make their journey upward. Trellises come in countless shapes, styles, and materials. You can easily make your own using simple bamboo poles secured at the top with twine to form a teepee. If you have a sunny exterior wall, you can grow herbs and shallow-rooted vegetables in wall-mounted planters, like repurposed rain gutters or even a hanging shoe rack.

Here are a few tips to help you make the most of your vertical garden:

- Make sure vertically grown plants are planted where they won't shade shorter, sun-loving plants.
- Grow plants on the south side of support structures to ensure they get the maximum amount of sunlight.
- Water frequently, since vertical crops dry faster than crops growing closer to the ground.
- Grow vertical plants in deep, well-draining soil so their roots can grow down into the soil rather than out, across, and just under the surface. Roots that grow laterally are more likely to crowd and compete with neighboring plants.

• Provide extra support for heavier fruit like melons and squash. Create slings made of old nylon stockings or netting to cradle young melons from the time they are a few inches in diameter until they're mature and ready to harvest.

EMPLOY SUCCESSION PLANTING.

Succession planting, sometimes called relay planting, involves planting different crops in the same plot, one after another, so the plot never lays fallow. As soon as one crop is harvested, a new one is planted in its place. This protects the soil from erosion and, if you choose crops properly, replenishes its nutrients. And with a new crop always on the way, there's more to eat.

Consider the following examples:

• **FIRST RADISHES, THEN TOMATOES.** Plant a quick-growing cool-season crop like radishes then, once harvested, plant warm-season transplants like tomatoes or peppers.

• **LETTUCE, THEN PEAS, THEN MORE LETTUCE.** Start with a spring crop of lettuce, then replace it with a short season summer crop of peas, then plant an autumn crop of lettuce.

• **PEPPERS FIRST, THEN LETTUCE.** After a late-summer harvest of peppers, sow an autumn planting of lettuce.

• **PEAS THEN BEANS.** Plant peas in early spring, then replace them with pole beans in late spring. Since both plants climb, you can use the same trellis to support both crops.

• **LETTUCE FOLLOWED BY CUCUMBERS.** Plant lettuce in early spring then, when temperatures climb, harvest the lettuce before it bolts and plant heat-loving cucumbers in its place.

• **BUSH BEANS, BUSH BEANS, BUSH BEANS!** Bush beans grow and produce quickly. As soon as you harvest one crop, start a new one. For variety, you can rotate different types of bush beans, such as green beans, pinto beans, black beans, and kidney beans.

• **FIRST PEAS, THEN COURGETTES.** After you harvest peas, sow or transplant courgettes or squash.

PRACTICE INTERPLANTING.

Interplanting is an intensive growing practice similar in concept to succession planting, except with interplanting you maximize yields by grouping two or more plants that grow well together but at different heights or at different rates.

One way to practice interplanting is to sow shorter crops around, in between, and under larger ones. For example, lettuce, greens, and parsley can be grown under and all around tomatoes and peppers. Another way to use interplanting is to plant fast-growing veggies among slower-growing ones so the faster crops can be harvested before they begin to compete with the slower ones. Many gardeners grow radishes, which can take 20 to 50 days to mature, among carrots, which mature in 70 to 80 days.

Succession planting and interplanting are great ways to increase your garden's yield. These methods don't work all that well in small containers, but I've had success using them in raised beds and large whiskey barrel planters.

GROW HIGH-YIELD CROPS.

Another efficient technique is to plant crops that have high yields per square foot. Tomatoes, peppers, beans, peas, courgettes, and cucumbers produce a lot of food from just a few plants. Plants like lettuce, kale, and Swiss chard are also great as they allow you to "cut-and-come again," meaning the outer leaves can be harvested regularly throughout the growing season while the plant continues to produce more leaves from its center. Crops like melons, pumpkins, and winter squash are not considered high-yield plants since they take up a lot of space relative to the size of their harvest, although you can get around that by growing them vertically. One-harvest crops like cabbage and broccoli are less ideal choices when space is limited.

GROW FAST-GROWING CROPS.

Some crops take a long time to mature, while others can be harvested just a few weeks after sowing them. Rocket, leaf lettuce, and spinach take 30 to 40 days, while beetroot, cucumbers, and bush beans are ready in about 50 days. Peas take 60 days, which is still pretty fast compared to most veggies.

PRIORITIZE EASY CROPS.

Sometimes it pays to be practical. Devoting too much time and soil to finicky vegetables can leave you with little to show for your efforts. Cauliflower, for example, is notoriously high maintenance. If you're just starting out, stick to staples, such as leaf lettuce, radishes, courgettes, kale, potatoes, green beans, peas, Swiss chard, summer squash, tomatoes, and cucumbers. Though not without their challenges, these crops are usually good bets for beginner and seasoned gardeners alike.

EXTEND YOUR GROWING SEASON.

When I first set out to grow my own food, I thought gardening was reserved for the warmest months of the year. Now that I've spent five years growing in a cold climate, I can assure you that is certainly not the case. In the natural world, plants finish producing flowers and begin dispersing seeds in the autumn. Left alone, Mother Nature sows these seeds soon after. We should follow her lead and sow seeds in the autumn too—especially since many early-season crops thrive in cold weather. Kale, beetroot, Brussels sprouts, and cabbage often taste better after a nip of frost. Others, such as alliums, actually need the winter to take root and grow. For everything else, the best way to extend the growing season is to protect crops with cold frames, hoop houses, greenhouses, row covers, and cloches (see page 155).

SELECT SHADE-LOVING VEGETABLES.

If your small space doesn't get a lot of light, fret not! There are many vegetables and herbs that can tolerate light to medium shade. Generally speaking, vegetables grown for their leaves and roots—leaf lettuce, spinach, rocket, beetroot, leeks, and potatoes—are the best choices for shady gardens. Many herbs tolerate partial shade as well, including basil, chives, lemon balm, mint, parsley, and chervil.

choosing a gardening method

Once you've assessed your space, you'll need to decide whether to grow directly in the ground or in raised rows, raised beds, or containers. Most gardeners use a mixture of several different methods. I grow most of my veggies in raised beds but grow herbs and flowers in the ground, in raised beds, and in containers. Here are descriptions of each method to help you get a better idea of which might be the best fit for you.

IN-GROUND BEDS

In-ground beds are the most traditional gardening method. Gardeners plant crops directly in the ground, after digging up and carting away existing vegetation, clearing out rocks and other debris, and breaking up hard or impacted soil.

PROS:
An in-ground bed is usually the cheapest way to start a garden. If you're experimenting with gardening for the first time, it's not a bad idea to start with a small in-ground bed to gain experience. An in-ground bed can be as big as the space you have, provided it's not paved over or otherwise inaccessible. It's generally easier to work with in-ground beds in odd-shaped or out-of-the-way plots, as you can dig them in any size and shape you choose. If you have rich, fertile soil, growing in the ground makes perfect sense. There's often no need to purchase soil or compost, although you will need to amend it after each growing season to replenish the nutrients in your soil.

CONS:
In-ground beds do not work as well if your soil quality is poor, and they should be avoided if there's any concern about soil contamination or pollution. If rock, sand, or clay predominate in your gardening space, it may not be possible to grow much of anything without long-term efforts to regenerate healthy soils. Loosening heavily compacted soil may require tilling, but excessive tilling can compromise soil structure and affect fertility (not to mention the backbreaking work involved!). Moreover, whatever the soil conditions, in-ground beds will be more vulnerable to pests and other diseases and often more accessible to wildlife and weeds. Some gardeners complain that in-ground beds are harder on their knees and backs, as they require most people to lean all the way over or kneel down to work in the bed.

RAISED ROWS

A raised row is a narrow mound of soil piled above the ground and mixed with compost and other organic materials, such as straw, leaves, and shredded leaves.

PROS:
Raised rows combine the simplicity of in-ground beds with the improved soil quality of raised beds. A raised row takes less time and effort to build than an in-ground bed and is less expensive than a framed raised bed. Since they're created by layering organic material on top of the ground, raised rows are particularly useful when the underlying soil quality is too poor to support a garden. Compared to in-ground

beds, raised rows are easier to maintain and tend to have fewer weeds. They also don't require tilling, which makes them easier on your back and better for soil structure than in-ground beds.

CONS:
A raised row garden often requires more space than a raised bed and might not look as clean or as organized as a garden with framed beds. Compared to raised beds, raised rows require more maintenance from year to year. Without a structure to hold the soil in place, the rows tend to erode faster, requiring gardeners to recreate and reshape pathways and rows every year. If you have back issues, raised rows might be a concern as well, as your only option is to squat or bend down to work close to the ground. With raised beds, you have the option to build tall bed frames that allow you to work at a comfortable height.

RAISED BEDS

A raised bed is a garden bed supported by a frame of wood, stone, bricks, metal, or concrete set on any flat surface—including turf, gravel, and concrete. Raised beds tend to be wider than raised rows, typically 2 to 4 feet wide (compared to 18 inches for raised rows). Although the up-front costs are higher, this method gives gardeners more control of soil, watering, and spacing.

PROS:
Raised beds present a number of advantages for gardens with limited space, since they can be custom-built to accommodate virtually any type of plot and can be planted at high density, resulting in more food per square foot. Raised beds also are great if you lack direct access to in-ground soil, as they can be placed on patios, decks, or even driveways. They also can protect your crops if you suspect your soil may be polluted or otherwise contaminated. Other containers can serve the same purpose, but raised beds are generally much larger. The soil in raised beds tends to warm faster than soil in the ground in the first weeks of spring, making it easier to get an early start on planting. Raised beds also make it easier to defend against rodents and other pests, especially if the beds are lined with hardware cloth.

CONS:
Cost is probably the biggest drawback to raised beds. They take time, effort, and materials to install; higher-quality woods and stone are especially expensive. Raised beds also must be filled with soil, compost, and other amendments. If you line the paths between raised beds with mulch or pea gravel, that adds to the overall cost as well, and that's before a single plant or seed has been purchased. And at some point, raised beds made of wood will need to be rebuilt or even replaced entirely as the wood siding wears out and rots away. There's a meaningful return for this investment, of course; raised beds are not a waste of money, by any means.

CONTAINER GARDENS

Any container can be a garden, or so I like to think. I use containers around my nine raised beds and elsewhere around the yard and inside my house. Big or small, grouped in little flocks or standing proudly alone, containers are great for many crops—especially those that might make unruly neighbors or those whose flowers add splashes of color to rows and beds of green. Containers also are highly mobile, letting you exploit changing light throughout the day and across the seasons.

PROS:
Containers give the gardener near total control. You can pick the size and shape of your little plot, depending on the containers you use. You can grow in intensive sunlight or full shade. And whatever you grow, containers make it easy to keep disease, pests, and weeds at bay. Plus, you can garden in the middle of winter with some containers in your home.

CONS:
Perhaps the biggest limitation to containers, aside from the obvious size constraints and the cost of procuring them, is the need to use potting soil. Potting soil is a mostly sterile medium, blended to allow for greater air and water circulation in the smaller, confined spaces of containers. Plants in potting soil generally need to be fertilized often and watered and drained carefully—not so much that they get waterlogged but not so little that they dry out and become root bound.

selecting materials for raised beds

If you choose to build framed raised beds, your first task will be to decide which material to use. There is a bevy of options to choose from and just as many factors to consider, such as safety, durability, sustainability, affordability, and aesthetics. Here are some of the pros and cons of the most common materials used to build raised beds.

REDWOOD OR CEDAR

Redwood and cedar are excellent options. They are easy to work with and naturally resistant to moisture, rot, and bugs. They also look beautiful and tend to last longer than other types of untreated wood. Redwood clocks in at around twenty years, while cedar typically lasts ten to fifteen years. The biggest downside is the cost. These are nearly always the most expensive types of wood.

When selecting any type of wood, look for boards marked "FSC"—which stands for the Forest Stewardship Council. FSC certification ensures that products come from responsibly managed forests that provide environmental, social, and economic benefits.

UNTREATED WOOD

Untreated wood is beautiful in the garden and less expensive than rot-resistant woods like cedar and redwood. On the down side, it tends to break down faster, although building with thick boards can make it last longer. Two-inch-thick larch, for example, can last up to ten years. One of the most affordable options is Douglas fir, which can last five to seven years.

BRICK OR CONCRETE

Brick is a beautiful material for raised beds and can last for generations. It can also be expensive, depending on the type of brick you use. Another option is poured concrete, which creates a clean look that can last longer than most other structures. Unfortunately, it also tends to be expensive and is hard to install without professional help.

Concrete building blocks are a cheaper alternative. They make for a durable, inexpensive building material that can be stacked three blocks high without needing mortar. If you go with this option, be sure to avoid the cinder block form of concrete blocks, especially older ones, as they may contain fly ash residue and traces of heavy metals like mercury, arsenic, and lead.

MORTARED STONE

People have built walls and beds from stacked stone for thousands of years. If built properly, a stone bed can last hundreds of years without maintenance. Whether cut from rock or organically shaped, mortared stone is a more permanent choice that adds formality to a garden. It is one of the least common types of beds as it can be difficult to DIY and expensive to pay someone else to do.

METAL

Metal stock tanks, generally used to feed farm animals, have become a popular option for raised beds in recent years. From an aesthetic point of view, they give the garden a clean, industrial look and, since they're already built, the only thing you have to do is add drainage holes to the bottom of each tank.

Stock tanks are fairly inexpensive and are readily available at most feed stores. They come in a variety of shapes and sizes but are typically round or rectangular with rounded ends. The biggest challenge with stock tanks is figuring out how to get them home from the store. They won't fit in the back of most cars, so unless you have a big pickup truck, you'll need to ask someone to help you. Although they've been galvanized to protect against rust, they're not completely resistant and can still rust over time. Another disadvantage of stock tanks is, even with drainage holes, they drain less efficiently than stone or wooden beds.

STEEL

Steel (or Cor-ten) is commonly used in landscape and construction projects and comes in sheets that are 3⁄16 or 1⁄4 inch thick. Although it starts out as a natural steel color, it weathers beautifully to a rusty patina. One of the advantages of steel is that it is naturally weather resistant and holds up well for many years. On the downside, installation is difficult and pricey. Steel containers also heat up easily and may not be the best choice for heat-sensitive plants.

MATERIALS TO AVOID

There are a few materials worth avoiding, for health reasons, when building raised beds for growing vegetables and herbs. These include railway sleepers, old tires, pallets, and pressure-treated wood, which may contain potentially harmful chemicals that could leach into the soil and, subsequently, your food. Unless you're using raised beds for growing inedible flowers or ornamental perennials, I would avoid these materials altogether.

If cost and sustainability are a concern, consider sourcing better, safer materials like fallen logs, willow, and hazel branches (which can be woven together), large stones, reclaimed bricks, or untreated, recycled boards.

choosing the right containers

When it comes to choosing the right containers for your garden, remember that anything that can hold soil can support a plant. Be creative and experiment with growing in whiskey barrels, buckets, sinks, grow bags, and troughs. As long as a container has drainage, doesn't leach harmful chemicals, and is big enough to support your plants, it's a perfectly suitable option. Personally, I think quirky little containers add charm and personality to a garden, and I have never been shy about upcycling an old sink or bathtub to grow some plants! As you go about collecting containers, consider these tips to help you make the best choice for your garden.

KNOW THE PROS AND CONS OF DIFFERENT KINDS OF CONTAINERS.

Terra-cotta and unglazed pots are beautiful, inexpensive, and easy to find, but they also wick away moisture from the soil and dry out quickly. Some plants, like rosemary or thyme, won't mind a dry environment; others will become stressed if you don't water them frequently. Because of their porosity, terra-cotta pots will also need to be brought indoors in below freezing conditions to prevent them from cracking.

Glazed pots, fiberglass, and plastic will retain moisture but may need to be given drainage holes if they don't come with them. Fiberglass and plastic pots are particularly appealing because they're lightweight and easy to move. Concrete containers are long-lasting and come in a variety of shapes and sizes, but are heavy and very hard to move. Because of their weight, they're not recommended for decks or balconies.

CHOOSE LARGER POTS.

It's easier to grow plants in larger containers than smaller ones. Larger pots hold more soil, which stays moist longer and prevents swift fluctuations in temperature. If plants are potted in small containers, they may need to be watered multiple times a day, especially in hot weather. To avoid having to "helicopter parent" your plants, choose larger pots or purchase self-watering containers that come with a reservoir for holding extra water.

EMPLOY HANGING BASKETS.

Hanging baskets make good use of vertical space. You can buy a traditional coconut coir–lined wire basket from your local nursery or upcycle an old colander, wicker basket, or watering can. Plants that grow well in hanging baskets include tomatoes, strawberries, lettuce, flowers, and herbs, such as basil, parsley, sage, and chives.

TRY USING GROW BAGS.

These flexible containers come in a variety of shapes, sizes, colors, and styles. Some bags are designed for specific crops; others are designed to be wall-mounted for vertical growing. Grow bags offer a variety of advantages. Most of them are made of breathable nonwoven fabrics that allow plants to shed excess heat and dry out between waterings. By the same token, they're almost impossible to overwater, which helps prevent waterlogging and root rot. Grow bags also promote healthy root growth and are easy to store and put away. The downsides are mostly related to durability. Depending on the quality of the bag, some types may not last more than two to three growing seasons, which can be hard on your wallet and the environment.

DON'T FORGET WOOD.

Wood is a wonderful material for container gardens. It's lightweight, durable, inexpensive, and biodegradable. It also blends well with a natural environment. You can buy wood containers premade or build structures with scrap wood. The drawback of wood is that it rots faster than other materials. You'll also need to make sure to use wood that has not been treated with chemicals.

CONSIDER PLASTIC.

For environmental reasons, plastic would not be my first choice, but if you can source it secondhand (or upcycle something you already own), it's a fine material for growing food. Plastic is durable, lightweight, and retains moisture. Just be sure to use a type of plastic that is safe for growing food. High-density polyethylene (HDPE), for example, is not known to transmit chemicals into soil or food. HDPE is marked with the number 2 inside the "chasing arrows" symbol you often see on plastic.

understanding soil

There's a saying in the gardening world that "there's an entire world beneath your feet." Not only is soil alive, but it's also one of the most biologically diverse habitats on earth. One shovelful of rich organic soil contains more species of microorganisms than the entire above ground Amazon rainforest.

Organic gardeners learn early on that soil is the foundation upon which everything in the garden is built. Healthy soil supports an abundance of healthy organisms, including bacteria, fungi, and earthworms, that together create a dynamic ecosystem. From good soils come plants that are stronger, more productive, and more resistant to pests and diseases. Although new gardeners want to learn how to grow healthy *plants,* experienced gardeners want to learn how to grow healthy *soil.* It's no surprise that a common mantra among gardeners is "Feed the soil, not the plant."

Growing healthy soil starts with understanding what kind of soil you have. This is fundamental to gardening in the ground, but less important when growing in raised beds or containers, where you have more freedom to mix your own soils. Regardless of which gardening method you use, you'll need to maintain your soil. Every plant you grow will draw nutrients from the soil, which must be replenished before the next growing season.

SOIL COMPONENTS

Though some gardeners are blessed with healthy soil, many of us garden in soil that is mediocre at best. Fortunately, it's not that difficult to turn poor soil into a plant-friendly medium if you understand what comprises healthy soil. Soil is composed of minerals, organic matter, air, and water. But the most important ingredient is the living organisms—the microbes, fungi, worms, insects, and small animals—which flourish when the components of soil are in balance.

Building a balanced soil starts with understanding each of these basic components:

MINERALS

About 45 to 50 percent of the volume of soil is made up of small pieces of weathered rock that have been broken down by wind, sleet, rain, and other natural processes. Soil type is generally classified by the size of mineral particles, going from clay (small) to silt (medium) to sand (large). The proportion of these particles in soil matters because it influences the texture of soil as well how well it drains and delivers nutrients to plants.

ORGANIC MATTER

Organic matter is nothing more than decayed plant and animal matter. Although it makes up just a fraction of the soil (around 5 to 10 percent), it is one of the most important components. Some gardeners refer to organic matter as the "soul of the soil," perhaps because it plays such a vital role in improving soil structure, retaining moisture, deterring disease, and feeding the organisms needed to create a balanced soil ecosystem. You can increase the amount of organic matter in your soil by amending it with compost, aged animal manure, leaf mold, cover crops, or mulch.

AIR

Healthy soil is about 25 percent air—which just so happens to be the amount insects, microbes, earthworms, and other soil life need to survive. All parts of a plant depend on air to breathe—the leaves, the stem, the roots, and even the flowers. The plant parts above the soil receive oxygen from the air through pores. The plant parts below the surface get oxygen from the air spaces in the soil.

Well-aerated soil has plenty of pore spaces between soil particles. To ensure there's a balanced supply of air in your soil, supplement it with lots of organic matter, avoid working with it when it is wet, and refrain from walking over your growing areas, which can compact soil and suffocate microbes and plants.

WATER

Another quarter of healthy soil is water, which is essential for growing plants and keeping microorganisms, worms, and other soil life hydrated. Water, like air, is held in the pore spaces between soil particles. One noteworthy fact about water is that it naturally gravitates toward smaller spaces, which explains why clay is so adept at holding it. The ideal soil has a mixture of small and large pore spaces so that it can hold both air and water well. The best way to improve your soil's ability to retain water is to supplement it with plenty of organic matter.

LIVING ORGANISMS

Healthy soil contains millions of living organisms and a robust soil food web. Soil organisms range in size from tiny one-celled bacteria, algae, fungi, and protozoa to larger earthworms, insects, and small invertebrates. These organisms support plant health by decomposing organic matter, cycling nutrients, improving soil structure, enhancing plant growth, and controlling pests and diseases.

SOIL TEXTURE

Gardeners love to talk about soil texture, which refers to the size of mineral particles in soil. As mentioned above, particles are classified by size from smallest to largest, as clay, silt, and sand. The larger the particle size, the larger the spaces, or pores, between particles. Nutrient-rich water moves through these pores, as does the air that microorganisms need to survive. When soil contains very little pore space, water and air can't move through it; when soil contains too much pore space, nutrients and minerals pass through too quickly.

For most plants, the ideal soil is 40 percent sand, 40 percent silt, and 20 percent clay. Soil with this makeup of particles is known as "loam" and provides the perfect amount of water-holding capacity, drainage, and fertility. Unfortunately, not all soils are loamy. Soils composed of too much of one type of particle can present challenges to gardeners.

CLAY SOILS have small particles that tend to pack tightly, leaving little room for water and nutrients to pass through it. Although clay is naturally fertile, it drains poorly, stays wet for a long time, and contains very little oxygen. Clay soil sticks together when wet, dries as hard as concrete, and is very difficult to cultivate.

SILTY SOILS have medium-size particles and pore spaces that hold some air and water. Silt is moderately fertile but, like clay, tends to pack tightly, especially when wet. It can be very powdery when dry and is easily carried away by runoff and wind.

SANDY SOILS are dominated by large particles, making them porous, fast-draining, and poor at holding water. Not surprisingly, sandy soils are usually deficient in nutrients.

LOAMY SOIL is a good combination of sand, silt, and clay particles, which allows it to hold water, air, and organic matter. Loamy soil supports good root development and holds on to the nutrients and water plants need to flourish. Loam is the goal.

TESTING YOUR SOIL

You can easily figure out what type of soil you have by doing a "squeeze test." Just take a handful of moist (but not soggy) soil from your garden, squeeze it in the palm of your hand, open your hand, and observe it:

- If it holds its shape at first but crumbles when you poke it, you have loamy soil (hooray!).
- If it holds its shape even when poked, you have clay soil. Clay soil will feel hard when dry, slippery when wet, and rubbery when moist.
- If it breaks apart as soon as you open your hand, you have sandy soil or silty soil. Sandy soil will feel gritty; silty soil will feel smooth, like flour.

TESTING SOIL FOR CONTAMINANTS

Soil can sometimes be contaminated with lead or other contaminants, especially in urban areas or near sites historically used for industry. If this is the case where you live, grow edible plants in raised beds or containers with new soil. Planting mixes certified by the Organic Materials Review Institute (OMRI) or U.S. Composting Council are free of heavy metal pollution. To find out whether your soil is contaminated, you can send a soil sample to a lab to be tested. Most of the companies that test for nutrients and organic matter will also test for heavy metals for an additional fee.

If you have less than ideal soil, don't be discouraged. Although loamy soil is ideal for gardening, you can still grow plants in clay and sandy soils. Adding organic matter can go a long way toward loosening clay soil and improving the water-holding capacity of sandy soil.

SOIL PH

Soil pH indicates the acidity or alkalinity of your soil and can have an impact on its ability to hold and supply nutrients to plants. The pH scale ranges from 0 to 14. A pH of 7 is neutral; a pH less than 7 is acidic; and a pH greater than 7 is alkaline. Most essential nutrients are soluble at pH levels of 6.5 to 6.8, which is why most plants grow best in this range. If soil is too acidic or too alkaline, nutrients can become chemically bound to soil particles, making them less available to plants.

To improve the fertility of your soil, it's worth the effort to ensure its pH is within the 6.5 to 6.8 range. While there are specific amendments that can address a pH imbalance (lime or wood ash can be added to raise the pH of acidic soils; sulfur can be added to lower the pH of alkaline soils), the easiest way to neutralize soil is to supplement it with organic matter, such as compost, leaf mulch, or aged manure.

You can test your soil's pH by purchasing a soil-testing kit from your local garden center or by sending a sample to a laboratory where it can be tested professionally. Many public universities offer mail-in soil-testing services, as do private companies. For in-ground gardens, raised rows, and raised beds, where you're adding topsoil or compost, it's a good idea to test your soil once a year. If you fill your raised beds with a commercial or homemade soil mix, you probably don't need to check it as frequently, unless your plants aren't growing as well as you'd expect.

SOIL FOR RAISED BEDS AND CONTAINERS

If you choose to garden in raised beds or containers, you'll need to buy soil to fill them. Choosing which type of soil to buy at the garden center can get a little confusing. The four types of soil most commonly available are topsoil, garden soil, raised bed soil, and potting soil.

TOPSOIL is soil (composed of clay, silt, sand, and organic matter) removed from the top 12 inches of earth during development projects. It's often used to fill holes, level ground, or as a component in garden soil. On its own, topsoil is a poor choice for filling raised beds—it contains organic matter but not enough to help plants grow to their full potential.

GARDEN SOIL is topsoil mixed with additional organic matter to improve its ability to nourish plants. By itself, garden soil is too dense to ensure sufficient aeration and drainage in raised beds and containers. It is typically used to amend existing soils or to create an optimal environment for in-ground gardening.

RAISED BED SOIL is the ideal choice for raised beds. It's rich with nutrients and drains well, balancing the best qualities of garden soil and potting soil. You can buy raised bed soil at a garden center or mix it yourself following the recipe on page 77.

POTTING SOIL is, ironically enough, a soilless medium, which is why it's often called "potting mix." It's made up of lightweight, fluffy, fast-draining, and moisture-retentive ingredients that provide a good environment for plants grown in pots or small containers. Most potting soils are a blend of peat moss or coconut coir (both hold moisture but drain freely), perlite (a lightweight, puffy volcanic mineral added to improve drainage), vermiculite (a mica-like mineral also added to improve drainage), and sand or compost.

Although you can certainly buy potting mix for your containers, you can also make it yourself, which saves money and gives you control over the ingredients you use. Taking this extra step might appeal to you if the blends you can find contain peat moss. Peat moss is problematic from a sustainability perspective because it is harvested from ecologically sensitive bogs and wetlands. Strip-mining peat destroys important habitat for plants and animals and releases carbon dioxide into the atmosphere. Instead of peat moss, I'd recommend using coconut coir, a renewable resource made from the outer husks of coconuts. Like peat, it holds moisture well, lightens soil, and promotes good air circulation.

To make your own potting mix, see page 84 for my favorite environmentally friendly recipe.

COMPOST

Soil and compost are not the same thing. While soil contains *some* organic matter, compost is 100 percent organic matter that has decomposed and is capable of feeding and conditioning soil. In addition to enriching soil, compost helps soil retain moisture and stimulates the growth of beneficial organisms.

TIPS FOR GARDEN COMPOSTING

You can buy compost at any garden center, but you can also make it yourself. If you've got the space, it's worth the effort. Not only will you save money, but you'll also divert food scraps and garden waste from the landfill to your garden. When our family started composting a decade ago, we were able to cut our household waste by more than 40 percent. If you want to give it a try, here are eight tips for setting up a successful garden composting system.

CHOOSE A COMPOST BIN OR CREATE A PILE.
A compost bin or an open pile will work. Bins have the advantage of keeping things neat and contained, while keeping out rodents and other critters. You can buy a bin or make one yourself.

CHOOSE A LOCATION FOR YOUR COMPOST.
The best locations are flat, well-draining, and easily accessible. In cool latitudes, place your compost bin or pile in a sunny spot with shelter to protect it from freezing cold winds. In warm, dry climates, place your compost in a shadier spot to keep it from drying out too much.

UNDERSTAND THE CONCEPT OF BROWNS AND GREENS.
There are two main ingredients in a compost pile: carbon-rich ingredients ("browns") and nitrogen-rich ingredients ("greens"). Carbon ingredients are the energy food for microorganisms and are typically dry, tough, or fibrous and tan or brown in color. The nitrogen group provide the protein-rich components that microorganisms need to grow and multiply. (See the chart on page 38 for examples of both types of ingredients.)

CREATE A SYSTEM FOR COLLECTING KITCHEN SCRAPS.
Have a system for collecting kitchen scraps ("greens"), whether you use a compost bucket, an old crockpot with a lid, or a large bowl you keep in the freezer to keep scraps from attracting fruit flies and smelling in the house. Refer to page 38 for a list of foods and materials that can and cannot be composted.

COLLECT AND STORE BROWN (CARBON-RICH) MATERIALS.
Create a system for collecting and storing brown materials to keep them dry, for example in a paper yard bag in a shed or garage or in a weatherproof aluminum trash can with a lid near the compost pile.

ALTERNATE LAYERS AND STRIKE A HEALTHY CARBON-TO-NITROGEN RATIO.
Create layers in your compost pile, starting with a 4-inch layer of twigs, hay, or straw, to allow for good air circulation, followed by a layer of dried leaves and a layer of finished compost. Then alternate between layers of green (nitrogen-rich) materials and brown (carbon-rich) materials.

Striking a healthy carbon-to-nitrogen ratio will speed up the process of decomposition. A good rule of thumb is to add four times as much carbon-rich ingredients as nitrogen-rich ingredients (by volume, not weight).

MAINTAIN YOUR COMPOST PILE OR BIN.
While you could leave your compost to decompose on its own, a little maintenance goes a long way to speed up the process. When you add fresh materials, be sure to mix them with the layers below it. Also, aim to keep compost the consistency of a wrung-out sponge—moist, but not too soggy. If it's too wet, add more brown materials; if it's too dry, add wet materials or water. If it's stinky, add a handful of shredded newspaper or straw. Use a pitchfork to mix or turn the compost once a week to introduce oxygen and eliminate odor.

USE THE FINISHED COMPOST.
It can take anywhere from two weeks to twelve months to produce compost. In general, you'll know it's ready when it's dark and crumbly, with a pleasant, earthy smell. At that point, you can use it to amend your soil or make compost tea (see page 122).

What and What Not to Compost

COMPOST ("GREENS")	COMPOST ("BROWNS")	DO NOT COMPOST
fruits and vegetables	corn cobs and stalks	black walnut tree leaves/twigs
eggshells	paper	coal or charcoal ash
coffee grounds	pine needles	meat or meat bones
coffee filters	sawdust	dairy products or eggs
tea bags/tea leaves	wood shavings	diseased plants
garden trimmings	straw	fat, grease, lard, or oil
grass clippings	dried vegetable stalks	cat feces, dog feces, or cat litter
house plants	dry leaves	pressure-treated wood
fresh leaves	hay	sand
hair and fur	nutshells	garden trimmings treated with chemicals
fresh manure	shredded newspaper	color or glossy paper
alfalfa meal	cardboard	weeds gone to seed
feathers	wood ash	

vermicomposting

If you have limited space for an outdoor compost bin, try composting indoors—with worms! Known as vermicomposting, this method uses worms to convert kitchen waste into a nutrient-rich organic matter euphemistically referred to as "worm castings." Not only are worm bins small and compact, but since there's no turning involved, they're also easier to maintain than compost bins. One of the best things about composting with worms is that it requires little in the way of time and money. To get started, all you need is a bin, newspapers, garden soil, and worms.

MATERIALS

worm bin (homemade or commercial)
50 pages of newspapers (black and white only)
spray bottle filled with water
garden soil
red wiggler worms

DIRECTIONS

1. Acquire or build a bin. I have a stackable wooden bin I purchased on Etsy, but you can make one yourself by drilling ¼-inch holes in the top, bottom, and sides of an opaque ten-gallon rubber storage bin. Some gardeners nest the bin inside a second bin so that the second bin acts as a catchment basin in the event that liquid drains from the first bin.

2. Composting worms live in moist bedding, which provides the air, soil, and water they need to survive. To prepare the bedding, tear the newspapers into ½-inch strips, then use them to fill the bin three-quarters full, keeping the bedding fluffy and not compact. Spray the bedding with water—it should be moist but not soggy, like a well-wrung sponge. Sprinkle 3 cups of soil on top of the newspaper to provide grit for the worms' bellies.

3. Add the worms. Not just any type of worm will do—you'll need red wigglers, which can be found at bait shops or specialty shops online. To figure out how many worms to buy, consider that one pound of worms will eat about half a pound of food scraps per day.

continues

4. Bury food scraps under the bedding. Before adding food scraps to your bin, cut or break them into 1-inch pieces. Then sprinkle them onto the surface of the bedding material and cover the scraps lightly with more newspaper. Most fruits and veggies are fine to use—just be sure to avoid citrus, meat, bones, oils, and dairy products.

5. Cover the bin. Place the lid on the bin and choose a spot with temperatures between 55°F and 80°F. Continue to feed, water, and fluff your bin and keep an eye on the worms to ensure they have enough food and moisture. Adjust feedings and waterings as needed, just as you would for a pet.

6. Harvest your castings. About two weeks before harvesting, start feeding your worms on one side of the bin only to encourage them to migrate there, so you can easily gather castings from the other side of the bin. If you have a multi-tray system like the ones sold online, put food in the tray above the one you want to harvest. In both types of systems, the worms will migrate to the food and away from the castings so that you can easily collect them.

MULCH

After compost, mulch is one of the best things to add to your garden. Broadly speaking, mulch refers to any material used to cover the soil's surface. Common organic mulches include leaf mulch, shredded bark, shredded leaves, straw, grass clippings, pine needles, and even compost. Among its many advantages, mulch helps control weeds, prevent disease, conserve moisture, maintain consistent soil temperatures, enrich the soil with organic matter, and keep the garden looking clean and tidy.

The best vegetable garden mulches are ones your property produces, including grass clippings, compost made from garden waste and kitchen scraps, and leaves collected from deciduous trees in the autumn. Grass clippings are particularly useful because they contain high levels of nitrogen and other important nutrients. If you do use grass clippings, be sure to collect them from gardens that have not been sprayed with chemicals.

You also can make your own leaf mulch. Just collect deciduous leaves in the autumn, shred them by running over them with a lawn mower, and store them in a brown paper bag until spring. Wood chips are a great option too, and they're easy to buy in bulk from local landscaping services. If you only need a little, you can also buy them by the bagful from your local gardening store. Avoid colored mulches, which sometimes contain harmful dyes that can contaminate your garden.

To use mulch, simply add it to your in-ground gardens, raised rows, raised beds, or containers twice a year in spring and autumn. Spread a thin layer on top—there's no need to work it into the soil.

TO TILL OR NOT TO TILL

While we're on the subject of soil health, it's probably a good time to talk about traditional gardens versus no-till gardens. Conventional wisdom believes that plowing, digging, and tilling is the best way to prepare soil for planting. Tilling is used to help turn up weeds, incorporate soil amendments, and create looser, fluffier soil for growing plants. If you've been gardening for a while, you've probably tilled or dug your soil.

Unfortunately, the process of "working" the soil creates problems as well. Over time, soil structure can become depleted and disrupted by repeated digging and tilling. Every time soil is disturbed, the network of organic materials and organisms is compromised. As beneficial organisms are brought to the surface, they dry out and die from exposure. Air pockets are eliminated and filled with crushed grains of tilled soil. Weed seeds buried within the soil are brought to the surface, where they find the light and air they need to germinate and grow. Although gardeners till to turn under existing weeds, this often exacerbates the problem and gives way to an explosion of new ones.

In recent years, many organic gardeners have begun using the no-till method of gardening. With this method, gardeners avoid intentionally disrupting the soil with plows, spades, shovels, rakes, hoes, and tillers and instead let worms, insects, and microbes cultivate it the way they do in nature.

The no-till method is full of compelling benefits—less digging, less weeding, higher yields, and healthier crops. The secret lies in regularly mulching the top layers of a garden with organic matter while leaving the lower layers undisturbed. Mulches cover the soil surface, protect it from erosion, smother and suppress weeds, and retain moisture. As the mulch decomposes, it enriches the soil and improves its structure, which is never compromised since there's no need to dig.

At the end of a growing season, no-till gardeners leave the roots of spent plants in place. Rather than yanking out a plant's root system, they cut the plant out at the soil line and allow its roots to decompose in the soil. The only downside to this method is that it takes time to build good soil, compared to the instant gratification of working in organic matter via tilling.

In the next chapter, on building, I'll outline one way to build a no-dig garden from scratch using what's known as the lasagna method (see page 80).

getting to know plants

Before you plan which plants to grow, it's helpful to have a good understanding of the ways gardeners distinguish different types of plants, understanding annuals, biennials, and perennials; warm-season crops and cool-season crops; and common plant families. It's also fun to delve into herbs, native plants, and edible flowers. For specific information on individual crops, hop over to "Plant Profiles," starting on page 231.

ANNUALS, BIENNIALS, AND PERENNIALS

Simply put, annuals are plants that perform their entire life cycle from seed to flower to seed within a single growing season. Perennials, on the other hand, persist for many growing seasons; in many cases, the top portion of the plant dies back in the winter and regrows from the roots the following spring. Biennials are plants that require two years to complete their life cycle so that they grow one season and bloom the next. This generally means you plant them and harvest their crops in year one and collect their seeds in year two. Knowing which plants are biennials is only important if you're interested in saving seeds. Otherwise, most gardeners grow them as annuals.

Most garden crops are annuals and will need to be replanted every year. Asparagus, rhubarb, and strawberries are a few of the most commonly planted edible perennials. There is also an array of perennial herbs, including chives, lemon balm, mint, rosemary, and thyme. Common biennials include beetroot, Brussels sprouts, cabbage, carrots, cauliflower, celery, kale, kohlrabi, leeks, onions, parsley, parsnips, spring greens, swedes, Swiss chard, and turnips.

WARM-SEASON VERSUS COOL-SEASON PLANTS

Annual vegetables generally fall into one of two groups: cool-season vegetables and warm-season vegetables. Knowing which ones prefer cool weather to warm weather (and vice versa) is invaluable when it comes to deciding what to plant.

Warm-season crops are the veggies of summer! They grow best in warm to hot weather and will not survive a frost. Generally speaking, warm-weather crops should be planted a couple of weeks after the last frost date of your area or when the soil is at least 60°F. Planting them indoors—or purchasing transplants from a nursery—can give you a head start on the growing season. Examples of warm-season crops include cucumbers, aubergines, melons, peppers, sweet corn, and tomatoes.

Cool-season crops grow best during cold weather and can be planted in spring or autumn. These plants don't just tolerate cooler weather—they need it to grow and mature properly. When warm weather arrives, many cool-season crops will bolt, or go to seed prematurely. Plants in this group can be planted when the soil and air temperature are at least 40°F, and flourish in temperatures below 70°F. Examples of cool-season crops include broccoli, cabbage, carrots, kale, peas, potatoes, and spinach.

Cool-season crops can be further subdivided into hardy or semi-hardy plants, depending on how well they tolerate cold temperatures. Hardy plants are the most adept at surviving cold temperatures, including freezes, frosts, and cold snaps. Their seeds will germinate in cold soils and will grow in daytime temperatures as low as 40°F. Seeds or transplants from this group can be planted as early as two to three weeks before the average last frost date in the spring.

Semi-hardy plants can endure limited or light frosts, meaning just an hour or two of frost or near freezing temperatures. They grow best when the minimum daytime temperature is between 40°F and 50°F. Semi-hardy veggies can also be sown as early as two weeks before the average last spring frost.

Some cool-season crops have a higher chance of survival when sown directly in the garden. Others can be started indoors and transplanted later. The chart below provides some guidance per type of cool-season crop.

Cool-Season Crops by Type and Growing Preferences

	HARDY VEGETABLES (CAN SURVIVE HEAVY FROST)	SEMI-HARDY VEGETABLES (CAN SURVIVE LIGHT FROST)
START INDOORS (THEN TRANSPLANT)	broccoli, Brussels sprouts, cabbage, leeks, onions, spring greens, swedes	artichoke, cauliflower, celery
DIRECT-SOW	kale, kohlrabi, peas, radishes, spinach, turnips	beetroot, carrots, endive, lettuce, potatoes, rocket, Swiss chard

PLANT FAMILIES

It can be overwhelming to try to learn about hundreds, if not thousands, of individual plants. Luckily, each plant belongs to a family, or group, that shares similar characteristics. Although each family has outliers, most of its members share similar physical characteristics, temperature preferences, growth habits, growing requirements, and pest and disease issues.

Once you know which family a plant belongs to, you can assume a lot about it. That might not sound all that useful, but practically speaking it means that if you know how to grow watermelons, for example, you'll probably have success with other members of its family, like cucumbers, pumpkins, squash, and cantaloupe. Knowing a plant's family can also help in planning for crop rotation. If you have a cabbage caterpillar infestation in your brassica bed one summer, you'll want to avoid planting members of the same family in the same bed the following summer. (Read more about crop rotation on page 58. Suffice to say, it's a good idea to rotate plant families from year to year to avoid problems with insects and diseases.)

There are hundreds of different plant families, but a gardener can do a lot with knowledge of just a few. Here I'll introduce you to the most common garden plant families and their characteristics.

ALLIUM FAMILY (ALLIACEAE): Also known as the onion family, alliums include chives, garlic, leeks, onions, shallots, and spring onions. They share a basic body type—grasslike leaves attached to a thick stem or bulb. Plants in this family are cold hardy and do the majority of their growing in the cooler months of the year. Alliums are heavy nitrogen users and like rich soil and plenty of water. They're fuss-free, easy to grow, and, compared to other garden vegetables, fairly resistant to pests and diseases. Chives and garlic are easy first alliums for new gardeners to grow.

BRASSICA FAMILY (BRASSICACEAE): Also known as the mustard family, brassicas include broccoli, broccoli rabe, Brussels sprouts, cabbage, cauliflower, cress, horseradish, kale, kohlrabi, mustard, radishes, rocket, spring greens, swedes, turnips, and watercress. Brassicas are classic spring and autumn crops. As cool-season plants, they do best when planted in spring before the last frost or again in late summer, for an after-the-last-frost harvest. Spring crops are usually started as seeds indoors and transplanted to the garden, whereas autumn crops can be grown from seeds sown directly in the garden in late summer. Not only can many of these plants survive frost, but low temperatures are known to enhance the flavor of some of them, like kale.

The leafy crops from this family are the easiest to grow, while cauliflower and broccoli can be quite finicky. It's best to avoid planting brassicas in the same spot year after year, as doing so can foster diseases that are particular to this family.

CARROT FAMILY (APIACEAE): Also known as Umbelliferae, members of this family have flowers that resemble an umbrella and include carrots, celery, chervil, coriander, dill, fennel, parsley, and parsnips. Most members of the carrot family are hardy, cool-season crops that can survive frosts easily.

Because most Apiaceae plants produce their edible parts underground, they grow best in loose, friable soil that drains freely. They also thrive when watered consistently, especially early on, before their seeds germinate. They are best sown directly outdoors in the garden six to eight weeks before the last frost date in the spring, and again in late August for an autumn crop. Proper thinning is essential for good production of these plants, as are good weeding and watering. Because of their ability to attract beneficial bugs and repel pests, plants in this family make great companions to many other plants.

CUCURBIT FAMILY (CUCURBITACEAE): Like alliums, members of the cucurbit family are easy to recognize. Not only do they all have large leaves on prickly, fuzzy stems, but none of them stand up on their own; instead, they climb using tendrils or creep across the ground. They're also some of the finickiest plants to grow and include squash, cucumbers, pumpkins, melons, and gourds.

As warm-season plants, cucurbits require consistent warmth, sun, and water to thrive. Regular watering, especially during early growth and as the plants blossom and bear fruit, is particularly important. Cucurbits like rich, fertile soil, enriched with organic matter, such as compost, aged manure, or leaf mold. In colder climates they should be started indoors and transplanted to the garden at the beginning of the warm season, so they have time to produce fruit well before the first frost.

In terms of special considerations, cucurbits often need a trellis to climb, or else plenty of space in the garden to spread out. They're also prone to pests and diseases, especially powdery mildew and squash vine borer.

ASTER FAMILY (ASTERACEAE): Also known as the daisy family, members include asters, dandelions, and sunflowers, as well as hundreds of varieties of lettuces and greens, such as romaine lettuce, iceberg lettuce, endive, escarole, and radicchio.

Although there's a lot of variety in this family, generally speaking, members require sunny locations with fertile soil, lots of organic matter, and consistent watering. Lettuces are my favorite members of this family and some of the easiest to grow. Not only do they grow well from seeds sown directly into the garden, but they also produce consistently throughout the growing season. As you harvest their outer leaves, they continue to grow new ones from their centers. They're ideal for small-space gardens because they produce prolifically without needing much room.

GOOSEFOOT FAMILY (AMARANTHACEAE): This cool-season-loving family includes beetroot, Swiss chard, quinoa, and spinach. Plants in this family can be grown throughout the spring and again in late summer. Swiss chard is a biennial and can live for two years in the garden as long as the temperatures aren't too extreme.

Members of this family can be harvested continually throughout the growing season. Although beetroot only produces one root, its leaves are edible (and highly nutritious) and can be harvested like those of Swiss chard. This family does well when sown directly from seed into the garden—especially beetroot, which doesn't like its roots disturbed.

Goosefoot crops are sensitive to heat, especially spinach, which will bolt (go to seed) in hot weather. To keep them from bolting, they may need to be planted in cool, shady areas in midsummer. Members of the goosefoot family like uniform temperature and humidity; mulching the top layer of the garden can help.

BEAN AND PEA FAMILY (FABACEAE): This family is made up of cool- and warm-season plants. Peas, broad beans, and fava beans are cool-season crops, whereas most other types of beans (green, lima, pole, etc.) are warm-season crops. Cool-season peas and beans can be planted as soon as the ground can be worked, sometimes well before the last frost date. Beans and peas are fairly low maintenance and easy to grow—just be sure to grow them in a sunny spot with well-drained soil.

Members of this family are known as nitrogen "fixers" because they make nitrogen available to other plants growing around or after them in the garden. They have a symbiotic relationship with soil-dwelling bacteria that allows them to convert atmospheric nitrogen to ammonium nitrogen, which is the only form of nitrogen that plants can use readily. Tomatoes, broccoli, peppers, and other common vegetables need nitrogen to grow but can't absorb atmospheric nitrogen. The only way they can get it is by absorbing ammonium nitrogen through their roots from the soil. For this reason, beans and peas make excellent companion plants for these crops.

NIGHTSHADE FAMILY (SOLANACEAE): If you hear the word *nightshade* and feel a little cautious, it's probably because you've heard they're poisonous. And though you're not wrong—this family does include some poisonous species, such as the deadly belladonna—the nightshade family also includes many beloved vegetables that are safe to eat, including tomatoes, peppers, aubergines, potatoes, and tomatillos. (However, it is important to note, the leaves of these plants are not edible.)

With the exception of potatoes, members of this family are warm-season crops that need warm soil and six to eight hours of direct sunlight a day. For best results, seeds should be started indoors six to ten weeks before the last frost date—that way, they can be transplanted as soon as the ground is warm, with a head start and ample time to grow to maturity during the growing season. Potatoes can be started outdoors from seed potatoes earlier in the spring. Nightshades need well-fertilized soil and benefit from a layer of mulch to help the soil retain moisture and maintain a consistent temperature.

Support is critical for all members of this family—tomatoes, aubergines, and peppers need structural support to hold their heavy fruits, while potatoes need to be "earthed up," or covered with loose organic soil periodically. This technique encourages potato tubers to grow deep and wide and allows new potatoes to form on top of maturing ones. Depth and darkness improve the flavor of potatoes; potatoes grown too close to the surface can receive too much sunlight and grow bitter.

HERBS

No kitchen garden is complete without herbs. They're easy to grow and maintain and can be used for cooking, steeping tea, creating remedies, and making crafts. Some gardeners use them to deter pests from—or enhance the flavor or productivity of—other plants. They also grow well in containers, making them ideal for beginner or small-space gardeners.

Herbs grow well just about anywhere, as long as they have full to partial sun and well-draining soil. Several herbs like coriander, parsley, chives, and mint will grow fine with as little as three to four hours of sunlight a day. Herbs like rosemary, bay, and lavender are native to the Mediterranean and will thrive in gritty, sharply draining soil.

Some herbs are low-maintenance perennials, such as rosemary, mint, lemon balm, oregano, sage, and thyme, and only need to be planted once in many hardiness zones. Others, like basil, coriander, dill, and parsley, can be planted yearly from seeds or starts. With so much versatility, herbs should be on every gardener's list when planning what to grow.

NATIVE PLANTS, WILDFLOWERS, AND COMPANION FLOWERS

Some gardeners are interested in growing more than vegetables and herbs—they want to create a wildlife-friendly garden that helps reestablish natural ecosystems and native species. In addition to being altruistic, thinking holistically about your garden is also practical: It is much easier to grow organically when you can enlist the help of native pollinators and beneficial insects. Many flowers make great companion plants to vegetables and herbs and even help keep pests at bay. Marigolds, for example, deter mosquitoes and rabbits, while chrysanthemums repel ticks and ants. Petunias add color to a garden while repelling asparagus beetles, leafhoppers, aphids, and tomato worms. Although it's hard to imagine anything not loving the smell of lavender, planting it around your garden is a great way to keep gnats and mosquitoes away.

The best way to attract bees, butterflies, and birds to your garden is to plant pollinator-friendly plants and flowers in a variety of colors, shapes, blooming periods, and heights. Native plants are particularly beneficial since they're suited to the local conditions and able to provide the ideal food source for native pollinators.

A pollinator-friendly garden can range from a decorative planter filled with native wildflowers to a large plot of land sown with pollinator-friendly plants. Our family planted two perennial pollinator gardens a few years ago and we fill several whiskey barrel planters with native wildflower seeds every spring.

EDIBLE FLOWERS. While we're on the subject of flowers, it feels imperative—critical, even!—to mention edible flowers. As much as I love growing veggies and herbs, there is something magical about growing flowers that can be used to add color and flavor to your favorite dishes.

Just be sure to do your research so you know which flowers are safe to eat and which are not. Also, while it's fine to eat edible flowers you've grown from seed, avoid buying them as transplants unless you can be certain they were grown organically. Commercially grown flowers are often sprayed with chemicals that make them unsafe to eat.

In most cases, you should eat only the petals, as the pistils and stamens can be bitter and contain pollen that can trigger an allergic reaction. Edible flowers can be enjoyed in salads, ice lollies, cookie batter, scone dough, and beverages.

EDIBLE FLOWERS

anise hyssop	daylily
bee balm	dianthus
borage	echinacea
calendula	honeysuckle
chamomile	lavender
chive blossoms	nasturtium
cornflowers	pansy
dandelion	viola

FRUITING TREES AND SHRUBS. Once you get the hang of growing vegetables and herbs, try adding berry bushes and fruit trees to the mix. Many fruits grow well in a home garden, even when space is limited. A grapevine can be grown over an arbor or a pergola. Lowbush blueberries and strawberries can be grown in raised beds. If you don't have space for full-size fruit trees, dwarf varieties are an excellent option. Many dwarf fruit trees and soft fruits (those that don't grow on trees) can grow in containers on a deck, patio, or balcony, or wherever you have space. Just make sure they receive adequate sunlight. Some of the best fruit trees for containers include dwarf varieties of apples, cherries, figs, peaches, nectarines, pears, plums, and lemons. The easiest soft fruits to grow in containers include blueberries, raspberries, blackberries, strawberries, and currants.

Before you run out and buy berry bushes or fruit trees for your garden, be sure to research which fruits grow best in your climate, and check that you have enough space and sunlight to support them. Keep in mind, if you grow in containers, you'll need pots large enough to support your plants as they grow, as well as potting soil and a good fertilizer to keep them healthy and happy. Also, be sure to check the best planting time, as most bare-root trees and bushes should be planted during the winter months, when they are dormant.

what to grow

Planting a beautiful garden takes time, effort, money, and maintenance. Starting out with a good garden plan can cut down on all of those things. It can take some patience up front, but setting aside time to plan will prevent heartbreak and headache in the long run. Once you've assessed your site, figured out what type of gardening method you plan to use, and learned how to differentiate between various types of plants, there are still other factors that should influence what plants you decide to grow. Take your time, have fun, and consider these general guidelines.

GROW WHAT YOU LIKE TO EAT.

Make a list of all the veggies and herbs (and maybe some fruits) you like to eat. If you need help, take a look at your grocery list—it should give you a good idea of what you consume most often. Consider also what you *don't* like to eat—if you don't like the texture of okra, for example, there's no reason to waste resources growing it in your garden.

CONSIDER WHY YOU WANT TO GROW FOOD.

If saving money is your top priority, try growing crops that are expensive to buy in the supermarket. If feeding your family organic foods motivates you, then grow crops that, when sold commercially, test highest for levels of pesticide residue. If your goal is to be more self-sufficient, grow highly productive vegetables, such as courgettes, radishes, tomatoes, salad greens, cucumbers, green beans, peppers, Swiss chard, and kale.

KEEP IT SIMPLE.

I know I just told you to make a list of everything you like to eat, but if you are just getting started gardening, consider a small, simple garden with just a few plants. If that sounds underwhelming, imagine how fulfilling it could be to grow a few tomato plants and lettuces. Not only does a small salad garden require minimal space, but greens grow easily and quickly, without much maintenance. A cut-and-come-again lettuce mix will produce multiple harvests in one season, and by late summer, you can enjoy your salads with fresh-grown tomatoes.

PLANT WHAT'S HARD TO FIND AT THE MARKET.

Commercial farming is all about maximum yields, transportability, and shelf life. Flavor is important too, but it's not a top priority. This means selection is often limited at the market. If you can never find your favorite unique variety of a vegetable at the market, consider growing it yourself. I have a hard time finding watercress, for example, and make a point every spring to set aside a large planter for growing it.

CONSIDER WHAT'S EASY TO GROW.

If you're a beginner gardener, you'll be rewarded and encouraged if you focus your efforts on plants that are easy to grow. Some of the easiest ones are basil, beans, chives, lettuce, oregano, parsley, peas, potatoes, radishes, rocket, spring onions, and Swiss chard. Wait to try challenging crops like aubergines, celeriac, and cabbage until you've gotten more comfortable in the garden and are ready to experiment.

PLANT WHAT TASTES BEST FRESH AND HOMEGROWN.

All crops taste best fresh and homegrown, but some can taste so sweet and full of flavor when eaten within a few hours of being harvested, you'll have a hard time going back to the store-bought variety. As far as I'm concerned, there's no comparison between store-bought and homegrown sweet corn, tomatoes, peas, and French beans.

PLAN ACCORDING TO HOW MUCH TIME YOU WANT TO SPEND IN YOUR GARDEN.

Vegetable gardens require constant attention. They need to be watered, weeded, fertilized, and harvested. You may need to manage pests and process the harvest. The size of the garden and the variety of plants dictate how much time you'll spend tending to it.

THINK ABOUT WHAT WILL GROW WELL IN YOUR CLIMATE.

Unfortunately, not everything you want to plant will grow well in your climate. If you live in a cold climate, skip heat-loving crops like tomatoes and cucumbers and put all your effort into cold-loving plants like spinach, lettuce, Swiss chard, and kale. If you're not sure what grows well in your climate (and microclimate), ask your neighbors or a member of staff at the local garden center what crops they swear by.

THINK ABOUT WHAT WILL GROW WELL IN YOUR SPACE.

Remember the guidelines for assessing your site, discussed on page 10, and think about what will grow well in your particular space. If your garden only gets four hours of sunlight a day, for example, focus on veggies and herbs that can tolerate shade.

THINK ABOUT SPACE AND DESIGN.

If aesthetics and/or maximizing your space are important to you, you may want to plant crops that can grow vertically (see page 16). Not only will planting up add some visual interest to the garden, but it will also help you optimize how much produce you're able to grow. Be sure to choose a few crops that can climb trellises, fences, or netting.

CONSIDER YOUR GARDEN TYPE.

Think about what type of gardening method you plan to use and choose your plants accordingly. If you plan to use containers only, for example, select plants and varieties that grow best in pots.

PRACTICE CROP ROTATION.

Avoid planting crops from the same family in the same spot year after year. Some pests and diseases that plague particular plant families persist in the soil. When you rotate same-family crops to different areas, you interrupt the pest and disease cycle. Crop rotation also helps maintain soil structure and nutrient levels, especially because some plants add nutrients to the soil while others deplete it. If you plant the same crop in the same place every year, the soil structure will slowly deteriorate. One example of crop rotation would be to plant corn and squash (which use a lot of nitrogen) in one bed one year, and beans (which add nitrogen to the soil) in the same bed the following year.

CONSIDER COMPANION PLANTING.

Companion planting is the practice of planting two crops in close proximity so they can enhance each other's health and productivity. In some companion relationships, one plant attracts beneficial insects to help pollinate a neighboring plant; in another situation, a fragrant herb repels pests from another plant. Low-growing plants can provide groundcover, while tall and climbing plants create shade. Although some plants make excellent companions, others are antagonistic toward one another. Plants that need and compete for the same resources, for example, usually don't make great neighbors. See the table on page 60 for a few plant pairings that are beneficial and some that should be avoided.

Common Companion Plants

PLANT	FRIENDS	FOES
beans	aubergines, beetroot, broccoli, cabbage, carrots, cauliflower, celery, cucumbers, peas, potatoes, radishes, squash, strawberries, tomatoes	garlic, onions, peppers, sunflowers
carrots	beans, lettuce, onions, parsley, peas, radishes, rosemary, sage, tomatoes	dill, fennel
corn	beans, cucumbers, lettuce, melons, peas, potatoes, squash	tomatoes
cucumbers	beans, cabbage, cauliflower, corn, lettuce, peas, radishes	aromatic herbs, melons, potatoes
lettuce	asparagus, aubergines, beetroot, Brussels sprouts, cabbage, carrots, corn, cucumbers, onions, peas, potatoes, radishes, spinach, strawberries, tomatoes	broccoli
peppers	basil, coriander, onions, spinach, tomatoes	beans, kohlrabi
radishes	basil, coriander, onions, spinach, tomatoes	kohlrabi
tomatoes	asparagus, basil, beans, carrots, celery, dill, lettuce, melons, onions, parsley, peppers, radishes, spinach, thyme	broccoli, Brussels sprouts, cabbage, cauliflower, corn, kale, potatoes

three sisters companion planting

One of the most well-known examples of companion planting comes from the Native American tradition of interplanting corn, beans, and squash in a trio called the Three Sisters. In this arrangement, the three plants grow symbiotically to enrich the soil, deter weeds, and provide support for one another. As the corn grows, beans find support by climbing up its stalks. In return, the beans fix nitrogen in the soil, which provides nutrition for the corn and squash. Meanwhile, the large, sprawling, and prickly leaves of the squash deter pests and block out weeds.

There are variations to the Three Sisters method. Many of these methods involve planting the corn, beans, and squash in clusters on low, wide mounds rather than in rows. Seeds are typically sown directly in the soil after the last spring frost when nighttime temperatures reach 55°F. Because corn has a long growing season, it's planted as early as possible to ensure the ears are ready to harvest before cold weather arrives. As for what kinds of seeds to use, you'll need pole beans, sweet corn, and summer or winter squash (pumpkins are unfortunately too vigorous and heavy!).

If you'd like to try this method, here are step-by-step directions.

1. Choose a location in full sun with rich, fertile soil that has been amended with plenty of organic matter.

2. Build a mound that is 12 inches tall and 3 feet wide at the base, with a flat top 10 inches wide. If you're building multiple mounds, space them 4 feet apart.

3. Sow four kernels of corn on the flat top of the mound, 6 inches apart and 1 to 3 inches deep, in a circle. Water regularly until the corn sprouts.

4. When the corn is about 4 inches tall, plant four beans halfway down the sides of the mound, at least 3 inches from the corn plant and 1 inch deep. Planting the beans a few weeks after the corn ensures that the corn stalks will be strong enough to support the fast-growing beans.

5. One week later, plant two squash seeds on either side of the base of the mound, approximately 18 inches apart.

100
120
140
160
60
180
40
20
°F
200
0
220

when to plant

When the first signs of spring appear, many gardeners tend to get restless. But spring can be fickle, especially in northern climates, sometimes vacillating from temperate to freezing temperatures from one day to the next. Experienced gardeners use a variety of methods to determine when to plant their crops, including observing frost dates, temperature, and nature. I like using a combination of all three to get the timing just right.

FROST DATES. A frost date is the average date of the last and first light freezes in the spring and autumn, respectively. Depending on a plant's tolerance to frost, it should be planted a certain number of weeks before or after these dates. Paying attention to these dates is important, because many plants will suffer or die if exposed to freezing temperatures. You can find local frost dates by calling your local cooperative extension service or by visiting one of many websites that provide dates according to zip code. Once you know your frost dates, you can refer to the directions on the back of seed packets to plan when to start seeds indoors or plant seeds and seedlings outdoors.

TEMPERATURE. Another way to decide when to plant crops is by taking the soil temperature. Frost dates are predicted based on historical averages, so they can vary quite a bit from year to year. Soil temperature readings, on the other hand, are specific to today's conditions. To take your soil temperature, insert a soil thermometer a couple inches into your soil and take a reading a few days in a row. Go slightly deeper if you are planting transplants instead of seeds. You can search online to find minimum soil temperatures for germination of most crops. Just keep in mind that an early warm spell that has warmed the soil doesn't necessarily mean all danger of frost has passed. If you are overly optimistic and plant too early, an unexpected cold snap could kill your seedlings.

NATURE. You can also take cues from nature to determine when to plant your crops. Phenology, as this study is called, looks at the timing of different events in nature and matches them with garden chores. Time-honored practices include planting parsnips, radishes, and spinach when crocuses start to bloom; planting semi-hardy vegetables like beetroot, carrots, and Swiss chard when daffodils begin to emerge; sowing lettuce, onions, and peas when forsythia starts to flower; and transplanting tomatoes when lily-of-the-valley has fully blossomed. Keeping a detailed journal of your natural environment will help you notice all sorts of patterns and cycles and connect with your garden in an attuned and personal way.

how to plant

Figuring out where to plant a garden is one thing. Deciding *how* to plant it is quite another, especially if you're striving for beauty and for practicality. When planning a garden layout, I like to factor in intensive growing practices such as vertical growing, succession planting, interplanting, and companion planting. I also think a lot about spacing—that is, how close or how far apart crops should be planted to maximize yields. Since my garden is my view from the kitchen, I care a lot about aesthetics too. If I'm going to look at my plot several times a day, I want it to be a source of pride and inspiration, rather than a trigger for what my grandmother called "mess stress." Whether your vegetable garden is big, small, neatly square, or all kinds of curvy; whether it's right outside your door or a little further away, here are a some things to consider when planning how to plant it.

SPACING

Most seed catalogs, plant tags, and seed packets give instructions for spacing in terms of two types of measurements: row spacing and plant spacing. Row spacing recommendations are based on growing plants in a traditional in-ground garden with long, single rows of vegetables spaced widely apart. The purpose of rows in traditional gardens is to give gardeners a path to walk between plants.

Since there are no rows in raised beds or containers, conventional spacing guidelines don't apply to them. Unless you're growing in an in-ground garden, you can ignore row spacing guidelines altogether and focus exclusively on plant spacing suggestions. For example, if you're planting peas and the instructions on the seed packet say to space seeds 2 inches apart in rows 18 inches apart, you can ignore the row recommendations and instead space seeds 2 inches apart on all sides.

If you're looking for a simple, reliable approach to plant spacing, try Mel Bartholomew's square foot gardening method. When I taught my children to garden, I found it the least overwhelming way to help them think through planning and planting. With square foot gardening, you divide a raised bed into a grid of 1-by-1-foot squares. So, a 4-by-4-foot raised bed is divided into 16 squares, and a 4-by-8-foot raised bed is divided into 32 squares. Each square is managed individually, with seeds or seedlings of each kind of crop planted in one or more squares at a density based on its size at maturity.

Square foot gardening spacing guidelines for common plants can be seen on the opposite page. You can figure out the spacing density for a plant by following these four steps:

1. Write down the plant spacing number on the back of a seed packet.
E.g., carrots, 3 inches apart

2. Divide the width of the planting section (1 foot = 12 inches) by the seed spacing number.
E.g., 12 inches/3 inches = 4 inches

3. Divide the length of the planting section (1 foot = 12 inches) by the seed spacing number.
E.g., 12 inches/3 inches = 4 inches

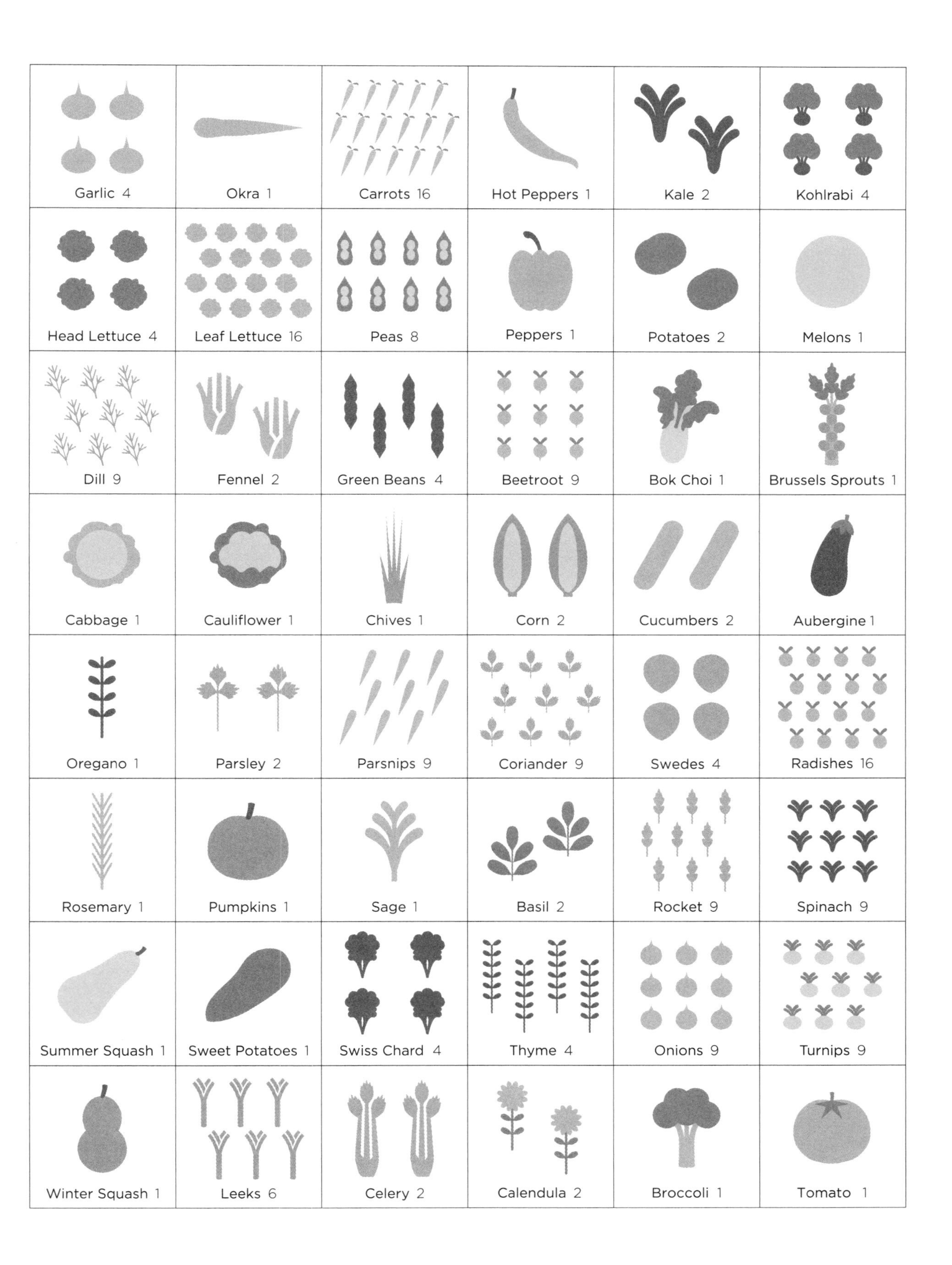
Garlic 4
Okra 1
Carrots 16
Hot Peppers 1
Kale 2
Kohlrabi 4
Head Lettuce 4
Leaf Lettuce 16
Peas 8
Peppers 1
Potatoes 2
Melons 1
Dill 9
Fennel 2
Green Beans 4
Beetroot 9
Bok Choi 1
Brussels Sprouts 1
Cabbage 1
Cauliflower 1
Chives 1
Corn 2
Cucumbers 2
Aubergine 1
Oregano 1
Parsley 2
Parsnips 9
Coriander 9
Swedes 4
Radishes 16
Rosemary 1
Pumpkins 1
Sage 1
Basil 2
Rocket 9
Spinach 9
Summer Squash 1
Sweet Potatoes 1
Swiss Chard 4
Thyme 4
Onions 9
Turnips 9
Winter Squash 1
Leeks 6
Celery 2
Calendula 2
Broccoli 1
Tomato 1

4. Multiply the two answers from steps 2 and 3 to determine how many plants can fit into each square foot.
E.g., 4 x 4 = 16 carrots per square foot

Some of the benefits of using square foot gardening are that it requires less weeding and results in higher yields in less space. It also takes a lot of the guesswork out of figuring out where to plant what. To create a square foot grid, you can shop online for ready-made structures that fit into standard size 4-by-4-foot or 4-by-8-foot beds or use twine and stakes to mark off squares yourself.

AESTHETICS

By definition, a kitchen garden is close by the house and in plain sight, making aesthetics an important consideration. Arranging plants by height is a good way to catch the eye—creating a sense of order and giving your garden a visually pleasing structure. The simplest way to do this is by following a short-to-tall format, placing the tallest plants toward the back of the garden and the shortest plants closer to the front.

Consider how you'll most often approach your garden when settling on which side is the front and which is the back. Or, if the garden is more or less equally accessible on all sides, you might try planting the tallest plants in the center of each bed or container and then sidestepping plantings outward from the center, with the shortest plants hanging along the edges of pots and beds. You can also beautify your garden by interplanting native plants, wildflowers, and edible flowers among your vegetables and herbs. I love to plant nasturtiums along the borders of raised beds so their broad leaves can drape over the edges and their flowers can add pops of orange color to the garden.

putting your plan on paper

There are few things I enjoy more in the middle of winter than spreading out a large sheet of paper on the kitchen table and sketching out a plan for my garden. I like to start a few weeks before it's time to order seeds and sow some of them indoors. My sketches are by no means works of art, but I take great satisfaction in creating a vision for what I want to plant and carefully rendering it on the paper before me. I use a pencil, of course, since planning is a process, and I always make a lot of changes along the way.

There are any number of ways to plan a garden, but I like to follow an adapted version of Jill McSheehy's step-by-step method from her book, *Vegetable Gardening for Beginners*:

1. Write down the average last spring frost date for your area.

2. Make a list of all the crops you would like to grow. Then use the guidelines for planning what to grow on page 56 to whittle your list down to crops you'll *actually* grow.

3. Divide the crops you'll grow into two groups: cold-season crops and warm-season crops.

4. Color-code each crop according to whether you plan to start seeds indoors (blue), purchase transplants from a nursery (green), or direct-sow seeds outdoors (brown).

5. Using the average last frost date for your area and the information on the seed packets, determine when to start seeds indoors (if applicable) and when to plant crops outdoors. Write the dates beside each crop.

6. Under each crop, write down its plant spacing requirements, how many plants fit in a square foot (if you're square foot gardening), days to harvest, plants that make good companions, and whether the plant needs vertical supports.

7. Sketch the borders of your garden on a piece of paper. If you like straight lines and right angles, graph paper is helpful, especially because each square of the graph paper can be a square foot of your garden. Whether you have in-ground beds, raised rows, or raised beds, try to draw the borders to scale to make planning easier. If you're using containers, include the size of the containers you plan to use so that you can determine the appropriate spacing for them as well.

8. If you plan to build or place vertical structures in your garden, include them in your sketch.

9. Fill the sketch of your garden with crops you would like to grow. Indicate how many plants you will grow in that space. If you plan to grow crops in succession, make a note of that. If this is not your first season growing in your garden, refer back to last year's plan so you can factor in crop rotation, if needed, and repeat anything that worked well.

10. Make changes as needed. Since you've used a pencil, it won't be hard to do!

keeping a garden journal

The best book about gardening is the one you write yourself. Nothing will better capture what's going on in your garden than the little notes, stories, sketches, and figures you jot in a personal garden journal. Journaling habits are as diverse as the gardeners who keep them. One gardener might scribble a note or two every week, another might keep careful tables teeming with data. Still others might record every little daily detail, and a few might fill each page with meticulously annotated drawings. Journals can be practical, poetic, quantitative, or just plain quirky. There are no rules for keeping one, so long as it reflects what is significant to you. Here are some suggestions for getting started:

- **GENERAL INFORMATION.** List your zone, frost date, and soil test results.

- **GARDEN HAPPENINGS FOR EVERY MONTH OF THE YEAR.** Each month, record what chores need to be done, what plants are in bloom, and any incidental observations you find useful. As the years pass, you'll likely add more tasks, reminders, and observations, for example, when to mulch, earth up potatoes, plant garlic, clean up the garden, and cover beds with row covers or sheeting.

- **PURCHASED SEEDS AND PLANTS.** If they're mail-order or online purchases, cut out the catalog photos and descriptions and paste them into your journal. If locally purchased or grown from seed, staple the plant tags or seed packs to your pages. Leave space to jot down notes on each plant, for example, whether it was robust, hearty, or flavorful.

GARDEN
CHORES

- **A SKETCH OF YOUR GARDEN LAYOUT.** Include sketches of your garden that show what you planted where. Not only will your sketch help you remember what you planted this year, but it'll also come in handy when you're looking back to plan what to plant next year. Detailed sketches are particularly useful when planning for crop rotation or when you want to remember where you planted autumn bulbs come spring.

- **A CHART TO RECORD THE PLANTS YOU ARE GROWING.** If you create a table, include columns to record the dates you sowed, hardened off, transplanted, and harvested each plant. A column to record special notes about each plant is also useful.

- **A RECORD OF PESTS AND DISEASES.** The same pests often appear at the same time each year. Write down pests encountered and the dates they appeared, to help alert you to potential infestations for upcoming years. Note what you did to minimize their damage, including anything you tried but didn't find successful.

- **FERTILIZERS YOU USED.** Record the type, amount, and date of any fertilizers you used to help you keep up with what works and doesn't work, as well as when you need to apply them again. A clear record can also help you troubleshoot growth problems.

- **PLANTS THAT NEED SPECIAL CARE.** If a plant requires annual pruning, deadheading, or staking, it's worth noting that so you are prepared and can perform the task at the appropriate time.

- **BIRD, POLLINATORS, AND OTHER WILDLIFE SIGHTINGS.** Just pretend you're keeping a bed-and-breakfast and having your animal guests sign in. Make note of all visitors, as well as when they came and what they liked to eat!

- **PLANTS RECEIVED FROM FELLOW GARDENERS, FRIENDS, OR FAMILY.** If you are gifted perennial plants, keep a record of who they were from and when you planted them. This information will be cherished by your children and grandchildren someday!

- **A "WISH LIST" OF PLANTS OR PROJECTS YOU'D LIKE TO TRY.** Paste in photos to remind and inspire you to try new plants, methods, and projects.

- **PHOTOGRAPHS.** Pictures are worth a thousand words—someday, you and your loved ones will relish looking back at the different ages and stages of your garden.

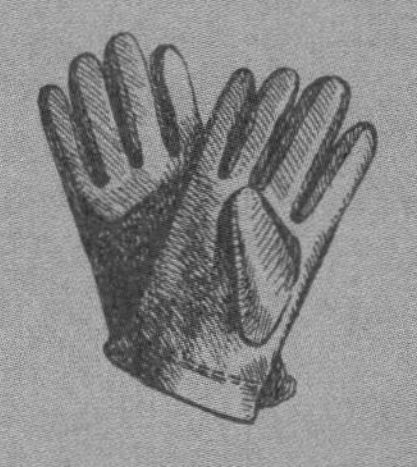

BUILDING

Once you've assessed your site, studied up on soil and plants, and made a plan, it's time to build your garden. There are many ways to do it—you could fill a library with books describing them all! My grandfather gardened in the ground; when I lived in Guinea my fellow growers used raised rows; my best friends in D.C. planted container gardens. I've done a little bit of all three, but my preference is for framed raised beds.

In this chapter, I'll show you two ways to create raised beds. The first requires clearing weeds and vegetation by hand. The second method, known as sheet mulching or lasagna gardening, removes weeds and surface vegetation by smothering them with layers (or "sheets") of organic matter. Because there's zero digging involved, it's better for the soil and easier on your back. It can be used for framed and freestanding raised beds, which makes it appealing and accessible to many types of gardeners.

I'll also show you how to prepare pots for a container garden and fill them with your very own potting mix. If you're hoping to extend your growing season, or simply don't have outdoor space, I'll also share tips for growing an indoor kitchen garden.

building a framed raised bed garden

Although planning and building your garden are the most labor-intensive stages of gardening, they're also the most important. If there's ever been a time to go slowly and think through details, this is it. After going through the trouble of building frames, schlepping compost, or setting up containers, you don't want to have to reverse your efforts and start over from scratch.

One spring, I hastily added a raised bed to my garden without taking the time to level it. Now it slumps to one side and creates an eyesore I can't seem to overlook. There's a part of me that would like start over, but there's also a part of me that can't find the time to do it! Needless to say, once a raised bed is full of soil, it's really hard to move it.

This section will walk you through a step-by-step process for creating a framed raised bed garden, from choosing a location to plotting your layout and building it to accommodate for pathways, borders, and other important structures. This is the process I used to create my fenced-in kitchen garden, and I could not be happier with the results!

If you'd prefer to use a no-dig approach that keeps the soil structure intact, skip to "Creating a No-Dig Raised Bed Garden" on page 80.

1. CHOOSE A LOCATION. The first step for building a framed raised bed garden is to choose its location. Head back to page 10 if you need a refresher on assessing your site based on factors such as sunlight, access to water, and convenience. Map out your property on a piece of graph paper, drawing your home, garden, patio, fences, sheds, and other existing structures, including existing trees or flower beds, to scale. Note the cardinal directions, label tap locations, and mark areas that get the most sunlight. If you have multiple site options, compare them against your priorities and goals. Remember, you can always create multiple gardens, or place beds in one area and containers in the other.

2. CHOOSE A LAYOUT FOR YOUR GARDEN. Once you've chosen a general location, design a layout. Decide on the overall dimensions and where you'll place beds, borders, pathways, climbing structures, compost bins, potting benches, and containers. There are countless ways to design a garden. In general you'll want to make sure it complements the style of your home, blends in with the existing landscape, and feels balanced and proportional to the eye. Think about what the garden will look like during different seasons of the year, for example, when it's peaking and full of foliage, fruit, and flowers midsummer versus when it has died back and is covered in snow and frost midwinter.

3. MEASURE THE GARDEN AND PLOT IT ON GRAPH PAPER. Once you have a general idea of the layout of your garden, plot it on a piece of graph paper so that each square on the paper represents a square foot of the garden. Carefully plan where to place beds and pathways. Once you create your border in the garden, it's hard to make changes without having to adjust the dimensions of your beds. If you plan to add other elements to your garden, such as containers or a potting bench, include them in your plan. When it comes to thinking about the dimensions and orientation of your beds and pathways, here are some details to consider:

PATHWAYS. Well-planned pathways will allow you to move easily and work efficiently. They should be wide enough to accommodate equipment, such as a wheelbarrow, but not so wide that you find it difficult to water between beds. Common wheelbarrows measure 2 to 3 feet wide, so a 3-foot pathway between beds is usually large enough for most gardens. I have eight beds inside my fenced-in garden, with four beds on each side of a 3-foot middle aisle. The rows between beds are narrower, at 2 feet wide, and just large enough for me to kneel or sit comfortably while sowing and weeding the garden. Depending on your level of comfort, you may need about the same amount of space or a little more.

RAISED BEDS come in all shapes and sizes. Generally speaking, it's best to keep them a maximum of 4 feet wide to allow you to reach the center without stepping on the soil. Beds against a wall or fence should be about 2 to 3 feet wide since you won't have access from all sides. Raised bed length is more of a personal choice and will depend a lot on the overall shape of your garden. Avoid making beds too long, lest you curse yourself for having to make long walks to get to the other side. For beginner gardeners, it's a good idea to start out with standard-size beds, such as 4 feet by 8 feet. Aim for beds that are at least 1 foot deep so they can accommodate leafy veggies, herbs, and root crops. Taller beds are great for gardeners with back or mobility issues. If you wish to garden while standing, beds that are 32 to 36 inches deep are ideal. You can also buy or build planter boxes on legs, which require less soil.

ORIENTATION is an important consideration. In midsummer, when the sun is directly overhead, you won't get much shade, assuming your site is unobstructed. But in spring and autumn, the sun is at an angle, and taller crops can cast a shadow on plants on their north side. Consider this when deciding whether the long end of your beds should be oriented from north to south or from east to west. Both can work, and sometimes shade isn't a bad thing (for example, when a tomato plant casts a shadow on a summer crop of lettuce), but you'll want to plan accordingly either way.

4. STAKE THE GARDEN. Once you've chosen a general layout on paper, mark off the garden by placing a stake at each corner of the plot. Wrap twine around the stakes to delineate the total area. If you want your garden to be square or rectangular shaped, make sure your corners are right angles and your borders are straight. You'll be removing vegetation in this area, so check that the marked footprint appears exactly how you want your garden to look.

5. CLEAR THE SURFACE VEGETATION. Remove surface vegetation, including grass and weeds, within the staked area you just created. To do this, first cut around the edges of the area with a sharp spade or edging tool. Then hold your spade at a low angle to the ground and use it to cut out blocks of vegetation. Peel the vegetation back with your spade and continue until you've removed all of it.

Once you've cleared the vegetation, use a rake to level the area. Then cover the entire area with cardboard or newspapers to slow the return of grass and weeds. If the weeds you removed haven't gone to seed, toss them in your compost pile.

6. MARK THE PERIMETER OF THE GARDEN AND ADD MULCH OR GRAVEL. Some gardeners like to define the perimeter of the garden with stone, brick, steel edging, or fencing. A 3-foot fence can help deter small rodents like rabbits, and a tall 8-foot fence can be helpful in areas where deer are a concern. Regardless of whether you edge around the outside of the garden, this is the time to stake out where you'll place the raised bed frames inside the plot. Before you set your frames in the garden, fill the garden space with 2 to 3 inches of gravel or mulch to prevent the plot from becoming muddy when it rains.

If you're trying to decide between mulch and gravel, there are pros and cons to both. Mulch is less expensive and easier to install than pea gravel, but it also needs to be replaced every year, as it decomposes quickly and washes away with rain. Mulch attracts insects and drains less effectively than gravel, but, on the flip side, it helps soil retain moisture and regulates its temperature.

7. ADD THE RAISED BED FRAMES. With your stakes in place, framing your beds should be fairly straightforward. Be sure to measure and double-check the distance between beds—and make any necessary adjustments—before filling them with soil. Check that your beds are level and, if they're not, add or remove mulch or gravel beneath the frames as needed. Level beds are important for many reasons. In addition to being pleasing to the eye, they prevent soil erosion during rainstorms or heavy watering.

8. ADD OTHER STRUCTURAL PIECES. If you suspect you may have a problem with subterranean rodents, consider adding a layer of chicken wire or other metal mesh to the bottom of your bed frames. The most durable and reliable option is galvanized hardware cloth with ½-inch squares. I do not recommend landscape fabric, as it provides no real benefits and hampers drainage. The great thing about raised beds is that the deep layers of soil and compost will smother most, if not all, undesirable growth. Any weeds that do rise to the surface will be easy to remove by hand.

9. FILL YOUR FRAMED RAISED BEDS. When it comes to filling raised beds with soil, you can use store-bought raised bed soil or make your own using any number of recipes. A simple, no-nonsense recipe combines 60 percent topsoil, 30 percent compost, and 10 percent potting soil. Garden soil and compost can be purchased by the bag from most garden stores.

If you are creating a large raised bed garden, you can save money and reduce waste by ordering in bulk from a local landscaping company. Look for local suppliers and ask to see a sample before you place an order. Remember the squeeze test from page 33? This would be a great time to use it! Bulk compost and topsoil are sold by the cubic yard. One cubic yard is equivalent to 27 cubic feet. To figure out how much soil you need, follow these simple steps:

- **STEP 1.** Multiply the length by width by depth (in feet) of each bed to determine its volume in cubic feet.
- **STEP 2.** Add the cubic feet of all of your beds to determine the total cubic feet.
- **STEP 3.** To determine the total cubic yards, divide the total cubic feet by 27.
- **STEP 4.** If you're using the above raised bed soil recipe, multiply the total number of cubic yards from step 3 by 0.6, 0.3, and 0.1 to determine how many cubic yards of topsoil, compost, and potting soil you need to order, respectively.

10. MAINTAIN THE SOIL WITH NO-TILL GARDENING. Although this method begins with some soil disruption when you remove the vegetation, once the beds are established you can practice no-till gardening. Instead of turning and digging soil to prepare your beds for planting each season, simply top them off with compost and other organic matter twice a year (once in the autumn and once in the spring). Remember that mulching is how you build healthy soil structure with no-till gardening.

My garden

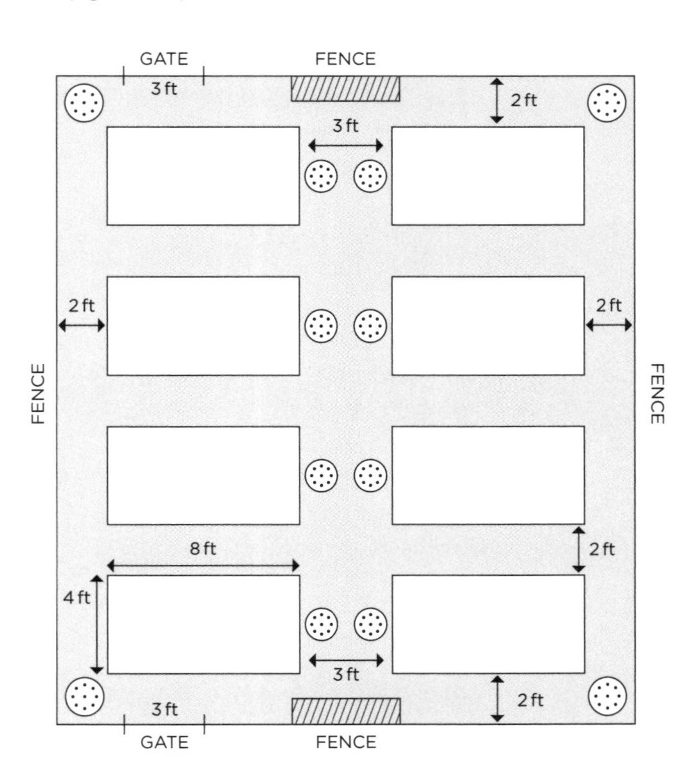

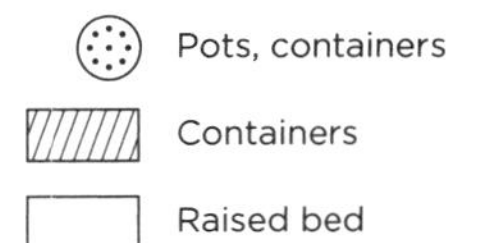

building a simple raised bed frame

Although you can buy raised bed kits online, it's not hard to build a frame from scratch. The following tutorial shows you how to make a simple wooden 4-by-8-foot frame that can be easily adjusted if you want longer or narrower beds. When purchasing lumber, consider the pros and cons of different types of wood as discussed on page 23, and inspect each board for straightness, as straight boards make for neater, tighter corners.

Note: This plan makes a 12-inch-high bed. To make a taller bed, you can stack additional 2x12 boards and adjust the height of the corner supports accordingly. You may also want to add a vertical center support to help secure the boards together.

You can adjust the length and width of the boards to make beds in any desired size; you may need to purchase longer or additional boards depending on your ultimate measurements.

MATERIALS	TOOLS
3, 2x12 boards, 8 feet long	**tape measure**
1, 2x4 board, 8 feet long	**hand saw**
about 26, 2½-inch galvanized deck screws	**power drill**
	screwdriver or hammer

DIRECTIONS

1. Cut one of the 2x12 boards in half to create two 4-foot-long boards. These will be the two short ends of the bed. The other two 8-foot-long boards will serve as the two long ends of the bed.

2. Cut the 2x4 board in half to create two 4-foot-long boards. You will use one 4-foot board as a center brace for the bed. Cut the other 4-foot-long board into four 1-foot-long pieces. These will serve as corner supports.

3. Set the boards into position. Drill pilot holes into the long board and then attach it to the short boards with screws. Repeat with the second long board.

4. Place one of the 1-foot-long 2x4 pieces in the interior corner between a short board and a long board and attach it to both boards with screws. Repeat for the other three corners.

5. Place the 4-foot-long 2x4 in the center of the bed between the two long sides. Align it with the bottom edge of the bed and check with the tape measure to ensure that it's centered and straight. Attach it with screws.

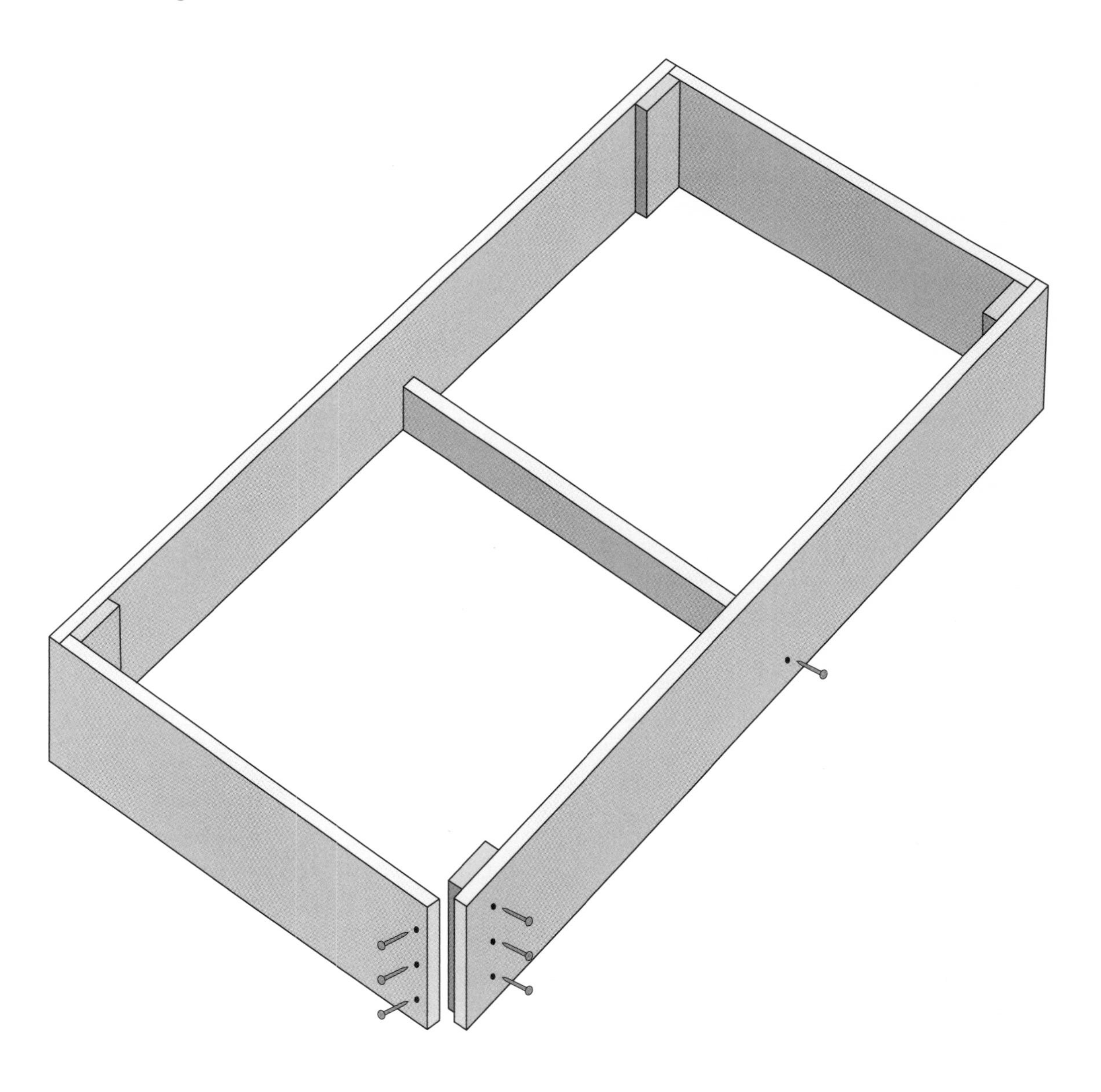

creating a no-dig raised bed garden

Whether for freestanding raised beds or rows or inside raised bed frames, many gardeners prefer to use no-till methods. (See page 43 for some of the advantages of no-till gardening.) One popular approach is the lasagna method, aptly named because of the way beds are built from layers of organic materials that "cook" over time and decompose into nutrient-rich soil. With the lasagna method, there's no digging or working of amendments into the soil—instead, you place several layers of organic matter on top of one another and let soil organisms break them down. As microbes, earthworms, and insects work their way to the surface, they dig and excrete more organic matter and build healthy soil structure in their wake.

The lasagna method mimics nature's way of building and enriching soil. Imagine a field where grass and other plants grow. At the end of the growing season, the plants dry out and fall down. As the trees surrounding the field lose their leaves, they blow onto the field and cover the first layer of plants. Animal droppings, nuts, and berries create another layer of organic matter. When rain and snow fall onto the ground, they add water to the layers, creating an ideal environment for them to decompose into healthy, fertile soil.

The lasagna method is one of the most cost-effective and environmentally friendly methods of gardening. You can build the layers from materials you already have, recycling yard waste, kitchen scraps, and newspapers back to the earth. The process looks a lot like composting, except instead of decomposing organic materials in a pile off to the side of the garden, you do it right on your garden beds. This method generally results in fewer weeds, thanks to the layers of newspaper and mulch that suppress them.

Autumn is the best time to start a lasagna garden. For one thing, the season provides easy access to leaves, straw, and other organic materials. Second, autumn rain and winter snow keep the materials in your lasagna garden moist, which helps them break down faster. Last, starting in the autumn gives the materials plenty of time to decompose before spring planting.

If you'd like to give it a try, here's a step-by-step guide.

1. **CHOOSE A LOCATION.** Follow steps 1 through 4 under "Building a Framed Raised Bed Garden" on page 74. Although you'll be creating beds (or rows) differently with this method, you'll still go through the same steps for choosing a location, deciding on a layout, measuring and plotting your garden, and staking your plot. Remember that you can build framed raised beds or freestanding raised rows with this approach. The choice is up to you.

2. **PREPARE THE GROUND.** Start by clearing the ground of any debris or rocks larger than an egg. If you are starting with a weedy plot or one with a lot of surface vegetation, cut or mow the weeds to ground level, rake up the trimmings, and remove them from the area. Never leave the trimmings in situ as they can provide a cozy habitat for slugs and other pests. You do not

need to remove vegetation or pull weeds, as they will be smothered by layers of organic material. If you have some lime and a nitrogen-rich organic fertilizer, such as blood meal or bonemeal, apply them to the area to help the grass decompose quickly once it's covered by newspapers and organic matter.

3. LAY DOWN NEWSPAPERS. Lay down newspapers in layers ½ inch thick over the entire area of your plot. Be sure to overlap the edges at least 6 inches to prevent grass or weeds from popping up. Avoid glossy printed paper or office paper, which contain potentially toxic inks and bleaches. Using a watering can or hose, water the newspapers, which will hold moisture and act as a weed barrier. The newspapers will break down over time as earthworms cultivate them.

4. MARK THE PERIMETER OF THE GARDEN. If you plan to create a border for your garden with stone, brick, steel edging, logs, or fencing, this is the time to do it. Use stakes and twine to ensure straight sides and square corners. This is also the time to stake out where you'll place the raised beds inside the plot. Regardless of whether you're creating framed or unframed beds, 4 by 8 feet is a good size. That said, the lasagna method can be used to create beds of any size or shape.

5. ADD RAISED BED FRAMES (OPTIONAL). If you're using raised bed frames, add the frames to your plot. Be sure to measure the distance between beds and make any necessary adjustments. Straighten the beds and get everything positioned exactly how you want it before moving on to the next step.

6. GATHER YOUR LAYERING MATERIALS. You'll need lots of "brown" materials and "green" materials to create the layers of your lasagna bed. Brown materials dry and decompose quickly and include dry leaves, shredded newspaper, straw, wood chips, sawdust, or newspaper. One word of caution: Do not use hay! Hay is often full of weed seeds that will sprout and compete with your crops. Green materials include compost, manure, untreated grass clippings, animal bedding, fresh yard waste, kitchen scraps, coffee grounds, seaweed, and blood meal.

7. LAYER YOUR BEDS. Start with green materials first and then alternate green and brown materials, being sure to water each layer. The brown layers should be roughly twice as thick as the green layers. Stop layering when your layers are 12 to 18 inches tall. Top them off with 3 to 5 inches of compost. If you're using framed raised beds, there's a good chance these layers will be higher than your frames. Don't worry! The layers will shrink significantly as the materials "cook" and decompose over the winter.

8. MARK PATHS. Put down thick cardboard to mark out paths between the beds. This will help suffocate and suppress weeds between the beds. Hold down the cardboard with stones or logs or cover the cardboard with 2 to 3 inches of wood chips or gravel. If you don't cover the cardboard with mulch or gravel, you may need to replace it once or twice before spring, depending on your weather conditions.

9. COOK THE GARDEN. You won't need to do a lot between autumn and spring, other than watch your garden "cook," shrink, and decompose. Just like with composting, the organic materials should be slightly moist but not so wet that they rot. In the case of an extended drought, lightly water to prevent the layers from drying out.

10. PLANT CROPS. Once spring arrives, it's time to plant! With this method, there's no digging or tilling needed to prepare your beds. Just hand pick any weeds that may have emerged and place your seeds and transplants directly into the newly decomposed soil.

11. MAINTAIN THE GARDEN. To maintain the garden, top off beds with a couple of new layers of compost and organic matter twice a year—once in the autumn and once in the spring.

preparing containers for a container garden

If you are creating a container garden or otherwise integrating containers into your space, you'll need to prepare your pots to provide the best possible environment for your plants. Although the process isn't complicated, it does require a bit more effort than simply dumping plants in a vat and covering them with soil. Pots that are being reused should be sterilized to prevent possible contamination from mold, fungi, or disease. Closed-bottom containers need to be given drainage holes to prevent plants from becoming waterlogged. Here are five steps to prepare containers for healthy, happy plants.

1. CLEAN THE CONTAINER. If you're using an old pot or have upcycled a used container, you'll need to clean and sterilize it. Remove dirt and debris from the pot, rinse it with soap and water, and spray the pot with a solution of hydrogen peroxide diluted with water at a ratio of 1:10. Scrub the pot clean and rinse it to remove traces of soap, water, and peroxide.

2. CREATE DRAINAGE HOLES. If your pot doesn't have holes, turn it upside down and drill one large hole in the bottom of a small pot or three to four holes in the bottom of a large pot. (You also may want to drill additional holes in a large pot if it only has one to start with.) Remove and discard drainage trays. After heavy rains, they will fill up with water and oversaturate your plants.

3. POSITION THE CONTAINER. Although your container may not be heavy when it's empty, it will be once it's full of soil. Position it where you want it to be in your garden before you fill it. If it's a large container you plan to move according to the sun, consider placing it on a wheeled plant caddy.

4. ADD POTTING SOIL. Fill your container with a potting mix formulated specifically for containers, leaving 1 inch of headspace for water and mulch.

5. PLANT, WATER, AND MULCH. Plant your seeds or transplants according to the directions on their packages and tags, then water the soil and top it off with 1 inch of mulch.

You may have noticed I didn't mention placing rocks or gravel in the bottom of the containers. Contrary to popular belief, doing so does not improve drainage and can often create the opposite effect.

FILLING YOUR CONTAINERS

If you only have a few containers to fill, buying ready-made potting mix from a garden store makes the most sense. But if you're building a large container garden, you can save a lot of money by creating a mix yourself. For an eco-friendly (peat-free) blend, I like to combine ingredients in a wheelbarrow, mix them well with a shovel, and store the mix in an airtight container until it's time to pot my plants. The recipe calls for sifted compost, which is simply compost that's been filtered to remove rocks, stones, sticks, and not-yet-fully-composted food scraps. If you buy ready-made compost it will have already been sifted.

ECO-FRIENDLY POTTING MIX FOR VEGGIES AND HERBS

Yields approximately 56 quarts

24 quarts sifted compost

16 quarts soaked coconut coir

8 quarts perlite

8 quarts vermiculite

2 cups vermicompost, if available

1 cup fertilizer specific to your plant's needs

ESTIMATING HOW MUCH POTTING MIX YOU'LL NEED

Potting mix is sold by volume, typically measured in quarts, whereas pots are measured in terms of diameter. If you're not sure how much potting soil to buy, here's a quick reference to help you make an educated guess.

CONTAINER SIZE	AMOUNT NEEDED
Pots and Tubs	
8-inch diameter	3 quarts
10-inch diameter	6 quarts
12-inch diameter	8 quarts
16-inch diameter	20 quarts
20-inch diameter	24 quarts
24-inch diameter	28 quarts
30-inch diameter	72 quarts
Hanging Baskets	
12-inch diameter	6 quarts
16-inch diameter	10 quarts
Window Boxes	
24 by 6 inches	12 quarts
36 by 6 inches	20 quarts

creating an indoor kitchen garden

Whether you want to extend your growing season or simply don't have space to garden outdoors, indoor kitchen gardening is a perfectly viable way to grow food. And although elaborate hydroponics systems, expensive indoor greenhouses, and other pieces of equipment are available, there's a lot you can grow with nothing more than a sunny, south-facing window and a set of florescent grow lights. If you want to give it a go, here are a few tips for creating a thriving indoor garden.

CHOOSE THE RIGHT SPOT. Since most plants need at least four to six hours of direct sunlight a day, a south-facing window is an ideal place to set up a garden. Be sure to choose a window that isn't obstructed by a tree or building that might create too much shade for your plants to grow. If you don't have enough natural light, invest in grow lights designed specifically for indoor gardening. For best results, aim to keep temperatures between 70°F and 75°F, and avoid windowsills near heating vents and cooking appliances.

ENSURE ADEQUATE AIRFLOW. Indoor edibles need air circulation to thrive. When airflow is impeded, the air can become damp and invite fungal diseases and pests. A cross breeze between windows works well but, if that's not an option, enlist the help of a small fan.

CHOOSE PLANTS THAT THRIVE INDOORS. Choosing plants is mostly a matter of space. If you're short on surface area, try growing basil, chives, dill, mint, oregano, parsley, savory, or thyme. If you have ample room, experiment with rosemary, sage, and lavender. For veggies, choose quick-maturing plants that you can grow in succession, such as microgreens, loose leaf lettuce, or wheatgrass. For herbs and veggies, you can start from seed or transplants, depending on time, space, and season (although some transplants can be hard to come by in the winter!). If you're growing microgreens or wheatgrass, you can grow a lot of food quickly from seed.

USE THE RIGHT CONTAINER. If starting with transplants, re-pot them in a container that is at least 6 to 12 inches deep. If planting herbs, plant several together in a wide or long container; otherwise, for individual herbs, use a pot that is at least 6 inches wide. If starting with seeds, use a smaller pot to begin, then transplant seedlings to a larger pot when they're 2 to 4 inches tall. If you're planting wheatgrass or microgreens, use a wide, shallow tray and sow seeds densely. Be sure to use containers with drainage holes and place a saucer beneath them to catch excess water and protect your windowsill from damage.

USE A SOILLESS POTTING MIX. Purchase and use a soilless seed-starting or potting soil mix formulated specifically for growing edibles indoors. Soilless mixtures use specific ingredients to optimize drainage, water retention, and air space, and they are free of disease and other contaminants. Regular soil, on the other hand, compresses over time, making it difficult for water to pass through it.

WATER AND SPRAY AS NEEDED. To test whether your plants need water, stick your finger in the top layer of soil. If it's dry, give your plants some water; if it's damp, let them be. After watering your plants, empty the saucers to eliminate excess water accumulation. Since indoor air can be dry, particularly in heated homes in the winter, fill a spray bottle with water and lightly mist the leaves and soil one or two times per week.

FERTILIZE AND TREAT FOR PESTS. Feed your plants a natural liquid fertilizer every few weeks. If your plants look like they're infested with aphids or other pests, spray them with diluted soapy water. I like to use a recipe of 1 tablespoon of castile soap diluted in 1 quart of water.

TRIM HERBS REGULARLY. Harvest veggies as needed and snip herbs often. Trimming herbs regularly will encourage them to grow full and bushy. Never trim more than one-third of the foliage at a time, as more than that can cause stress. To harvest microgreens and wheatgrass, never pull them out of the soil; instead, trim them at soil level with sharp scissors.

ARTIFICIAL LIGHTS. If you want to germinate seedlings or grow high-light-requiring edibles such as basil and cherry tomatoes, you may need to increase your light levels with artificial grow lamps. This is especially true in winter months, when the days are shorter and natural light levels are lower.

PLANTING

Once you've planned and built your garden, you finally get to do what you set out to do in the first place: grow plants! Gardening may seem pretty straightforward—put some seeds in soil, give them a drink of water, make sure they have light, and wait for them to grow—but there's a bit more to getting started than you might expect. As with planning a garden, planting a garden requires forethought. And a good garden often begins indoors, in the heart of winter.

starting with seeds or seedlings

Before you grow your crops, you'll need to decide how to start them. Generally speaking, there are two ways to do it—with seeds, which you can start indoors or outdoors, or with seedlings, which are also known as transplants or starts.

Growing from seed has a lot of advantages. Seeds cost less than transplants, with a packet of one hundred tomato seeds typically costing less than a single tomato plant. The bigger your garden, the more you'll save by growing from seed. Growing from seed also lends itself to greater variety. Seed catalogs feature dozens, if not hundreds, of different types of crops while plant nurseries typically carry only a few common varieties. If you're looking to grow rare, historic, or heirloom veggies, starting from seed is often the only way to go.

While growing from seed has its advantages, some gardeners still prefer starting with transplants. They're faster and often easier to grow. Some plants struggle to grow from seed, and getting it right can involve trial and error and invite delays. Some can be direct-sown outdoors, but generally not until the weather is warm enough (the exact optimum conditions will vary by plant). Other plants need to be sown indoors, which takes more effort than some gardeners are willing to make. The truth is, unless you've got a greenhouse or a grow light system inside your home, it's usually less of a hassle to buy plants as starts.

PLANTS TO DIRECT-SOW	PLANTS TO TRANSPLANT	PLANTS TO DIRECT-SOW OR TRANSPLANT	
beetroot	artichokes	basil	kohlrabi
carrots	asparagus	beans	leeks
coriander	aubergines	bok choy	lettuce
corn	celeriac	broccoli	melons
dill	celery	Brussels sprouts	mustard greens
garlic	mint	cabbage	okra
lettuce, baby mix	peppers	cauliflower	parsley
onions	rhubarb	chives	pumpkins
parsnips	rosemary	cucumbers	spring greens
peas	sage	dill	spring onions
potatoes	strawberries	endive	squash
radishes	sweet potatoes	fennel	summer squash
rocket	thyme	kale	Swiss chard
swedes	tomatoes		tomatillos
spinach			winter squash
turnips			

Even if you do have a system, however, there are a few plants you *shouldn't* start indoors. Plants like beetroot, carrots, peas, and spinach don't like their roots disturbed and instead do better when they start and finish in the same place. With these plants, it's best to wait until the weather is right to direct-sow them in the garden.

SOURCING SEEDS

There are any number of ways to find seeds for your garden. Garden shops and even some grocery stores have racks of seeds for sale. Seed swaps with friends, neighbors, and other local gardeners can be a great way to discover new varieties. And seed catalogs and online seed suppliers give you access to hundreds of different types of seeds, including those best suited for your local conditions. I prefer to shop for seeds online, usually by midwinter, not long after I've worked out a plan for my garden. I can take my time looking at the options and can get exactly what I want and in the quantities I need.

I sometimes pick up a packet or two of seeds at garden shops, but I always make sure the seeds are organic, viable (check the expiration date), and from a reputable company. Seed savers take note: If you're planning to save seeds from the plants you grow, make sure the original seed packet indicates that they are open pollinated (OP), not a hybrid (see page 166 for more information about saving seeds).

And I do love finding out about seed swaps in our little community. Gardeners have been trading seeds since the first crops were raised. In some cultures, seed swaps are a hallmark of annual harvest celebrations. Whether you're trading a favorite herb or a treasured family heirloom, seed swaps are a great opportunity to connect with local gardeners, glean tried and true gardening tips, and diversify your seed supply.

TESTING SEED VIABILITY

If you have leftover seed packets from a previous season, don't just throw them away! First, test the seeds to see if they're still viable. Set five to ten seeds on a damp paper towel. Fold the towel in half, covering the seeds, and place it in a storage container with an airtight lid. Place the container in a warm location, for example on top of the refrigerator, and watch the seeds over the next couple of weeks to see if they germinate. The percentage of seeds that sprout will give you a good idea of whether the remaining seeds in the packet are worth planting.

STARTING SEEDS

If you decide to grow from seed, keep in mind that you can start them inside your home, in small pots or trays, or outside directly in the garden. Starting indoors means you can begin early, well ahead of the growing season. This can make a big difference in cooler climates, where the growing season starts late and ends early. Transplants sold in nurseries were started indoors, usually offsite in large greenhouses.

In terms of timing, many seeds started indoors are sown four to eight weeks before their prescribed transplantation date. But it's best to check, as some seeds need to start earlier or later than others. The seed package should tell you the best time to start, with guidance for starting indoors versus outdoors, based on your area's last expected frost date.

Sowing seeds indoors can be a bit of a project—you'll need a growing medium, containers, warmth, light, moisture, and time for feeding, thinning, repotting (if needed), and hardening off. In the waning days of winter, it's not uncommon for the homes of ambitious gardeners to look like a runaway science project. Here's what you'll need to grow seeds indoors (and see page 99 for a step-by-step tutorial):

GROWING MEDIUM. Seeds grown indoors should be started in a soilless seed-starting mix composed of fine particles, moisture-retaining amendments, and sterile ingredients that won't stimulate fungal or bacterial growth. Seed-starting mixes are different from potting mixes, which are coarser in texture, less adept at draining, and amended with nutrients and fertilizers. Seeds benefit from finer soil which allows their roots to develop properly. They don't need nutrients since—not unlike eggs—they carry inside them all the nutrients they need to sprout. Once the seeds sprout and form true leaves (the third and fourth leaves to develop), they'll need to be repotted into a container with potting soil to keep them growing strong until it's time to transplant them into the garden.

CONTAINERS. You can start seeds in anything from seed trays or modules to used yogurt containers, so long as there are holes in the bottom for drainage. Although I typically prefer to use biodegradable supplies, plastic six-packs and flats are suitable choices that can be reused year after year. You can also make seedling pots from recycled toilet paper rolls or newspapers (see the tutorial on page 179), either of which can be directly transplanted into the garden. Avoid terra-cotta pots and compressed peat pellets, as they dry out too quickly and risk overstressing young seedlings (although I have used terra cotta with success as long as I watered frequently). If you are reusing containers from seasons past, be sure to clean them beforehand to ensure there are no lingering fungi or diseases.

WARMTH. Seed-starting happens in two different stages: germination and growing. Germination is the sprouting stage, when the root and leaves emerge from the seed. For most plants, light isn't needed during this stage because it happens under the soil, but you'll need a warm (but not blazing hot) environment. You can buy heat mats to slip under trays or place your trays on top of a refrigerator, which is often surprisingly warm to the touch. If you prefer to upcycle something you already own, a great gardening hack is to place incandescent (not LED) outdoor Christmas rope lights under your flats to warm the soil.

LIGHT. Once the seeds have sprouted and leaves have emerged, the sprouts will need a lot of light—ideally, fourteen to sixteen hours per day. In many climates, the late-winter and early-spring days are still quite short, so while a sunny window is always a good idea, it usually helps to add a fluorescent light, suspended just an inch or two over the tops of the plants. You'll

need to raise the lights as the seedlings grow, eventually maintaining a distance of 4 to 6 inches from the tops of the plants.

MOISTURE. Keeping soil consistently moist, without overwatering plants, is key to the health of young seedlings. Avoid letting soil fluctuate between the two extremes of bone dry and sopping wet and instead aim to maintain a sponge-damp consistency. Use a spray bottle to mist the soil from above, or water seedlings from below by placing plants in a tray filled with 1 inch of water so the soil can absorb it naturally through capillary action. Don't let pots sit in water too long—twenty to thirty minutes is usually plenty of time.

FEEDING. When a seed sprouts, the first leaves to unfold are called cotyledons. All the nutrients that are needed to sustain the plant at this stage are contained right inside the seed. But once the second set of leaves ("true" leaves, which have a different shape than cotyledons) appear, you'll need to begin feeding seedlings an organic liquid fertilizer, such as liquid kelp or fish fertilizer (following the manufacturer's directions), until you transplant them into the garden.

THINNING. Depending on your germination rates, you may have several seedlings growing in each pot or cell. As hard as it may be, you'll need to thin them out, removing the weakest seedlings to ensure the strongest ones have room to grow. A good rule of thumb is to leave one seedling per 3-inch pot and two seedlings for larger containers. To thin the seedlings, use a pair of scissors to cut them off at the soil line. (If you have a lot of thinnings and can't bear to compost or throw them away, toss them in a salad and enjoy the sprouts for lunch!) Never pull the seedlings out of the soil, as you could damage the delicate roots of neighboring plants.

REPOTTING. Some seedlings may outgrow their pots before it's time to transplant them into the garden. These plants will need larger containers in order to continue growing at a healthy rate. If you're not sure whether a plant has outgrown its pot, observe how often it needs water and check to see whether its roots have filled the container. The thirstier it is and the more crowded the roots look, the more likely the plant needs more room to grow.

STEP-BY-STEP GUIDE TO REPOTTING

1. To prevent transplant shock, water the seedlings well before repotting them.

2. Fill new, larger containers with pre-moistened potting soil up to ½ inch below the rim.

3. To gently remove the seedling from its original container, invert it and squeeze the sides. Place one hand under the container with the main stem of the seedling between your index and middle fingers and shake or tap the container to loosen the soil.

4. In the new pot, dig a hole that's the same depth as your plant and about twice as wide as the root mass.

5. Don't bury your plant too deeply—cover it just up to its neck or the spot where the root and the stem meet. The one exception to this rule is when planting tomatoes, in which case you

should dig a deeper hole and try to bury as much of the main stem as possible.

6. Gently center the seedling in its new container, fill in the gaps with potting soil, and tamp it in lightly.

7. Water the repotted transplant well, label the new container, and return the plant to the window or lights.

HARDENING OFF. About a week or so before it's time to transplant your seedlings to the garden, you'll need to begin acclimating them to outdoor conditions. This transition period is called hardening off and involves setting seedlings outside for increasingly long periods of time. The exposure to cooler temperatures and outdoor air prepares them for life in the garden and reduces transplant shock.

To harden off your plants, start by setting them outside in the shade for three to four hours. Bring the plants back indoors at night. Repeat this process every day, increasing the time they spend outside by one to two hours per day, gradually exposing them to morning sun, and eventually leaving them outside overnight. After seven to ten days, your plants will be ready to move to their new home in the garden.

SEEDS

starting seeds indoors

When deciding how many plants to sow, start with more than you think you will need. Some seeds will fail to germinate, and some seedlings will die before they make it to the garden.

MATERIALS

organic seed-starting mix	seeds
water	pencil
trowel	spray bottle
growing containers	humidity dome
waterproof tray	heat mat (optional)

DIRECTIONS

1. Place the seed-starting mix in a large bowl. Add water and stir the mixture with a trowel. Continue to add water until the mixture is slightly damp but not soaking wet.

2. Fill your containers with seed-starting mix up to ½ inch below the rim and gently but firmly press to remove air pockets. Place the containers on the waterproof tray.

3. Check the seed packet instructions to see how deep to sow your seeds. Small seeds can often be sprinkled right on the surface; large seeds need to be buried to about two to three times as deep as the seed's thickness. To plant larger seeds, use a pencil to poke a hole into the soil in the center of each container, at the depth indicated on the seed packet, then drop in the seed and cover it with the soil mixture. Press gently. Instead of covering fine or tiny seeds, press them into the seed starting mix. After sowing the seeds, mist the surface.

4. Label the containers with the plant variety and sowing date. (Wooden plant markers or craft sticks hold up well and can be composted when it's time to discard them.) Cover the containers with a humidity dome to seal in moisture and place the tray in a warm spot—on a heat mat or the top of the refrigerator.

continues

starting seeds indoors, continued

5. Check your seeds daily, mist them with water if the soil surface is dry, and wait for seedlings to emerge. The estimated timing for germination varies by plant and depends on the temperature; expected timing is usually given on the seed packet.

6. Once the seeds sprout, remove the humidity dome and place the tray under full-spectrum florescent grow lights, placed 2 inches from the tops of the seedlings.

7. Water plants with a spray bottle or from below to keep the soil moist during germination and growth.

8. Once the second set of leaves ("true" leaves) form, begin feeding your plants an organic fertilizer, following the instructions on the package.

9. Thin the weakest plants by cutting them off at soil level. Leave the strongest plants to grow.

10. Harden off the seedlings, then transplant them into the garden.

NOTE: *If you notice mildew in or around your plants, increase air circulation by removing the humidity dome. Use a spoon to gently scrape mildew off the soil without damaging your plants.*

transplanting seedlings

Starting your garden with transplants has many advantages. Transplants give you a head start on the growing season and more control and predictable results in the garden. Whether you purchase them from a nursery or start them from seed at home, transplants are garden-ready as soon as it's warm enough to put them in the ground. This means they'll mature sooner and give you an earlier harvest. Starting early can make or break a crop if you live in an area with short summers (like I do!). In the case of long-season crops like melons, tomatoes, and peppers, using transplants helps ensure your crops mature before cold weather returns.

Whether you start your seedlings at home or buy them from a nursery, you'll need to harden them off before transplanting them into the garden. You'll also need to check the last frost date to make sure it's safe to transplant them. Many garden centers sell plants several weeks before it's safe to transplant them outdoors, so it's always a good idea to double-check.

And a word of warning: Even if the calendar says it's time to plant, your local weather conditions might not agree. Be sure to observe the short- and long-term forecast so you can be as certain as possible that the last spring frost has passed. When it's finally safe to transplant your seedlings, wait until early in the morning on a warm, overcast day so your seedlings can settle into their new homes before you expose them to intense midday sun. If cool, wet weather is in the forecast, wait another day or two until the weather is suitable.

Once you've found a good time to transplant your seedlings, follow these simple steps to move them from pot to soil.

1. Water your seedlings well. Then, while they're still in their pots, set them on top of the soil where you intend to plant them or mark the spots with craft sticks or stones.

2. Remove each seedling from its pot by inverting it and squeezing the sides of its container. Place one hand over the container with the main stem between your fingers and shake or tap the container to loosen the soil.

3. Dig a hole that's the same depth as your plant and about twice as wide as the root mass. Don't bury your plant too deeply; instead, plant it up to the neck of the plant, or the spot where the root and the stem meet.

4. Gently center the seedling into the hole and backfill it with the soil you removed from the hole. If you're transplanting your plant into a pot, fill the container with soil up to ½ inch below the rim of the pot to accommodate watering. Water the transplant well and label it, if needed.

Don't worry if your seedlings look droopy or unhappy on the first day. It takes new transplants a few days to settle in to their new surroundings, and even longer to put energy into new growth. If the temperature drops unexpectedly or you're concerned about pests like slugs and snails, cover your seedlings with an old cotton sheet or row cover to give them added protection. You can read more about row covers on page 155.

direct-sowing seeds

Unlike starting seeds indoors, sowing seeds directly outdoors exposes them to unpredictable elements such as weather, wildlife, and insects. Even so, many plants dislike being transplanted and will fare better if you sow them in a bed or container where they can germinate and grow without being uprooted or disturbed.

When it comes to direct-sowing seeds, timing is everything. Soil temperature is a crucial factor for deciding when to sow seeds outdoors. Direct-sowing is easiest in mild climates with early springs and late winters with above-freezing temperatures and reliable precipitation. In less forgiving climates, direct-sowing is still an option; you just have to pay more attention to timing. If your growing season is short, you'll need to plant seeds as soon as it's safe to do so and choose crops that will have enough time to mature before cold weather returns in the autumn. Fortunately, most seed packets provide specific information about when to sow seeds, as well as how many days it takes a particular variety to go from seed to harvest.

Although direct-sowing requires less preparation than starting seeds indoors, you can't just cast seeds over soil and expect good results. A few of your seeds may sprout, if you're lucky, but you'll be much more satisfied with your harvest if you follow these best practices for sowing seeds outdoors.

- **ENRICH THE SOIL.** Prepare your beds by adding a generous amount of compost to the soil several weeks before planting. By the time you're ready to sow seeds or transplant seedlings, soil organisms will have worked the nutrients into the soil. You can also stimulate the growth of beneficial bacteria (and suppress weeds) by adding a layer of cardboard or biodegradable egg cartons under the compost.

- **REMOVE WEEDS.** Remove all weeds from your beds and containers so you can tell the difference between a seedling and a weed once the seedlings start to grow. Otherwise, when you go to weed, you might pluck out your baby rocket plants instead of dandelion sprouts.

- **LEVEL BEDS.** Use a rake to gently level your beds and remove roots, rocks, and other large objects that could impede the growth of tiny seedlings.

- **SOW SEEDS AT THE RIGHT TIME.** If your seed packet says to direct-sow seeds after the last frost has passed, don't risk it and plant your seeds before that date. Spring is fickle, and a week in the 80s is no guarantee that it will stay that way. Some seedlings won't survive a frost if the temperature suddenly dips toward freezing.

- **PLANT SEEDS.** Follow the directions on your seed packet for how to plant your seeds. Large seeds should be planted deeply to ensure good seed to soil contact, adequate moisture, and ample darkness. Small seeds usually need to be planted close to the soil surface so that sunlight can trigger germination. If you want to plant seeds in a straight line, use a piece of string—anchored with stakes on either side of your beds—as a guide.

- **WATER GENTLY.** Moisten the ground where you've planted your seeds, but be careful not to overwater, lest you wash away your seeds. After planting, be sure to keep the soil evenly moist while waiting for the seeds to germinate. Once the plants emerge, water them whenever the soil looks dry.

• **THIN SEEDLINGS AS NEEDED.** Just as with sowing seeds indoors, you'll need to thin your outdoor seeds to give the strongest seedlings room to grow. Use small scissors to snip weaklings or pinch them out with your forefinger and thumb.

• **MULCH.** After the plants have germinated, emerged from the ground, and grown strong stems, add a 1- to 2-inch layer of fine mulch around the plants. Mulch inhibits water evaporation and slows weed growth, but too much can limit the flow of air and water and stresses young seedlings.

TENDING

A garden happens in stages. Once you get to tending, you can let out a deep breath and wipe the sweat from your brow (and probably leave smudges of potting soil all over your face!). There's still plenty of hard work ahead, but now you get to enjoy the fruits of the labor you put into planning, building, and planting. I love tending my garden—it's second only to harvesting—because it revolves around the nurturing work of taking care of plants.

This section explains the most essential tasks for keeping your garden healthy, including watering, weeding, feeding, supporting, pruning, hand-pollinating, preventing pests and diseases, and extending the growing season. But first, you'll need some tools!

helpful tools

You don't need a shed full of tools to care for a garden. In fact, your hands can do a lot of the work. From sowing seeds to spreading compost, applying mulch, pinching suckers, and removing dead leaves, a bit of dexterity goes a long way. That said, there are a few helpful tools that are essential to maintaining a garden. Rather than buying them all at once, consider acquiring them on an as-needed basis; if you buy high-quality tools and take good care of them, they can last for many years. Tools with stainless-steel heads tend to be especially long-lasting.

Choose handheld tools that fit comfortably in your hands. If you have children in your family, pick up a few child-size tools to make it easier for them to work alongside you in the garden. Most hardware stores and nurseries carry small-scale sets for preschoolers and slightly larger, more functional sets for school-aged children.

GARDEN GLOVES. Although I always forget to use them, gardening gloves do a lot more than keep your nails clean. They help prevent scratches from thorns and branches and protect your hands from blisters, splinters, and irritation from handling certain plants. A heavy-duty pair is good for pruning, digging, and moving material; a lightweight pair, coated with rubber, is great for helping your grip and keeping your hands warm and dry.

GARDEN HOSE. Water is the foundation of life for plants, so it's important to have a hose that can reach and spray every area of the garden. Measure the distance between your tap and the farthest point of your garden so you can be sure to purchase one that is the right length. A hose nozzle or watering wand gives you extra reach and is also handy for customizing the spray pattern and water pressure. When not in use, be sure to keep your hose out of direct sunlight to prevent it from melting or cracking.

HAND CULTIVATOR. Sometimes called gardening claws, hand cultivators are like miniature rakes. They're helpful for loosening and preparing soil for seeds and for removing debris around the base of plants.

HAND TROWEL. A basic tool, a hand trowel is great for transplanting seedlings, planting containers, and removing weeds. A trowel with a broad blade helps move more soil at once, while one with a narrow blade is particularly suited for digging up weeds.

PITCHFORK. A pitchfork is one of the best tools for turning, moving, and harvesting compost, as well as for lifting and pitching organic materials such as straw or leaves.

PRUNING SHEARS. Also called hand pruners or secateurs, pruning shears are essential

for pruning vegetables, herbs, and fruit trees. They are especially useful for pruning vining tomatoes, harvesting thick-stemmed vegetables such as cabbage, and snipping woody herbs like rosemary. A good pair of kitchen shears are invaluable for harvesting cauliflower, broccoli, and squash as well as soft herbs such as basil, coriander, dill, thyme, and parsley.

RAIN GAUGE. There are different types of rain gauges, but a simple analog model—usually mounted on a post or stuck in the ground with a metal pole stake—is all you need to keep track of how much water your garden receives each week.

RAKE. A rake comes in handy during autumn and winter cleanup when leaves have fallen on the beds and pathways of your garden. There's no need to go out and buy anything special; a standard leaf rake will do. Just make sure the tines are sturdy and won't break if you use it for other types of work around the garden.

SPADE. These square shovels are workhorses in the garden. They make easy work of digging holes for plants, shoveling compost, and moving materials from one area of the garden to another. They come with long and short handles; longer ones provide more leverage but also are heavier.

STAKES. Stakes are the most basic way of supporting vine tomatoes and climbing plants such as beans, peas, and squashes. Bamboo or wooden stakes are nice-looking, but metal will last longer. Fallen branches can be used if you're on a budget or prefer a more natural look.

TWINE. Twine is useful for tying stakes together to make teepees or for providing beans with something to climb. It can also be used as a marker for planting seeds in a straight line.

WATERING CAN. A watering can isn't strictly necessary but might come in handy for smaller gardens or when enlisting the help of children in the garden. Galvanized cans last longer than plastic ones and won't rust. Consider the size of the can relative to your (or your child's) strength—one gallon of water weighs more than eight pounds!

WHEELBARROW. A wheelbarrow is a great tool for hauling compost, gravel, or mulch from one part of your property to another. A good wheelbarrow should be strong, sturdy, and well balanced, so you can haul a full load without it tipping. Be sure to buy one that can fit between your beds (or design your beds to fit a wheelbarrow!). Store it clean and dry to prevent rust.

watering

For plants, water makes the world go round, and when it comes to vegetables, it's easy to understand why. Many veggies are almost completely made up of water, with cucumbers and lettuce comprised of 96 percent water, radishes and celery made up of 95 percent water, and peppers and spinach consisting of 92 percent water. In addition to filling their tissues, plants use water to carry out germination, photosynthesis, nutrient transfer, and transpiration.

Although watering your crops may seem like a no-brainer—just turn on the tap and spray your hose—there are a surprising number of ways to muck it up. If you water plants too little, they'll get stressed and grow shallow roots; if you water them too much, you'll suffocate their roots. Never mind the pests and diseases you'll attract by watering their foliage at the wrong time of day! Luckily, when it comes to getting it right, the learning curve is not that steep. Here are a few guidelines to help you water your plants efficiently and effectively.

AMEND YOUR SOIL. You can prevent a lot of watering issues by amending your soil with heaps of organic matter. Compost holds six times its weight in water. Organic matter also provides food for soil organisms, resulting in improved soil structure. The better the soil structure, the more effectively soil can hold and transport water from wet areas to dry areas. If you're gardening in containers, however, you'll need to be careful not to add too much organic matter, or your soil may become compacted.

MULCH, MULCH, MULCH. Evaporation is a natural process, and although you can't prevent it altogether, you can certainly slow it down. To help soil stay cool and moist, top it off with leaf mulch, shredded bark, shredded leaves, straw, grass clippings, pine needles, or compost. Mulching is not just for beds—containers benefit from it too.

GIVE PLANTS ROUGHLY ONE INCH OF WATER PER WEEK. Most plants and herbs thrive on about 1 inch of water per week, which can come from rain or a hose. A simple rain gauge can help you measure how much rain your garden receives, which will in turn help you estimate how much supplemental watering you need to do. One inch of rain should soak to a depth of 4 to 12 inches, depending on the composition of your soil.

WATER DEEPLY RATHER THAN FREQUENTLY. A common mistake many new gardeners make is watering their gardens often and lightly, which unfortunately results in shallow, poorly developed roots that tend to dry out and cause plants to wilt and underproduce. A better way is to water less frequently and more deeply, with the goal of moistening the soil 6 to 8 inches deep. You can test the depth of water penetration with a moisture monitor. Watering well encourages a plant's roots to grow deeply into the ground, where they're better able to withstand stressors like heat, wind, and drought.

WATER THE SOIL, NOT THE FOLIAGE. When using a hose or watering can, aim the water toward the base of plants rather than at its leaves. Water that collects on leaves can promote disease, especially for plants like tomatoes, beans, peas, melons, and squash. You could also opt for an irrigation system, such as soaker hoses or drip emitters. Both systems release water slowly into the soil and can be fitted with a timer to automate watering.

WATER EARLY IN THE MORNING. Watering late in the day can cause foliage to stay wet during the night, which attracts insects, slugs, fungi, and diseases to your garden. A better time to water is early in the morning when it is still cool and water can reach a plant's roots before it evaporates. Watering early also gives leaves plenty of time to dry before nightfall.

CHOOSE THE RIGHT CONTAINERS. Clay pots, such as terra cotta, are highly porous and wick moisture from the soil, which can be a good or bad thing depending on the type of plants you're trying to grow. Metal pots heat up quickly and cause potting mixes to dry out. If you want to water your containers less often, consider using large pots made from recycled plastic or glazed clay. If you don't like the look of plastic, try nesting it into a more decorative pot (with drainage!). If you don't have time to water pots daily, consider investing in self-watering containers. These pots allow you to fill a reservoir with water once a week; the soil wicks it up and delivers it to the roots as needed.

WATER SEEDS AND SEEDLINGS WELL. Seeds need to stay continuously moist from the time they are sown until the time they germinate and emerge from the soil. For small seeds, such as carrots, celery, and most of the cabbage family, it's important to keep the soil moist at least 1 inch deep. To prevent the soil from drying out during this critical period, you may need to water once or twice a day and/or use a humidity dome to slow evaporation. For larger seeds, such as beans and peas, it's okay if there is some dryness at the soil surface—just be sure to keep the seedbed evenly moist. Give special attention to seedlings, too, as they are more vulnerable to drought than well-established plants.

USE WATER CONSERVATION METHODS. Although your plants don't care where their water comes from, for environmental and financial reasons, it's a good idea to conserve resources by collecting rainwater in barrels or even reusing water from sinks or baths, especially if you live in a dry climate. Rain barrels are often set up to collect rain running off a roof into a gutter and downspout. Most barrels hold 60 gallons of water and are fitted with a valve for filling watering cans or feeding a drip irrigation system. You don't have to spend a lot of money to collect rainwater. Secondhand containers can function as barrels or tanks—you'll just need to connect them to a downpipe with plumbing pipes. If you plan to collect and reuse sink or bath water, use soap made from natural ingredients that won't harm your plants.

Another way to conserve water is to borrow the ancient technique of using ollas to water crops. These unglazed ceramic pots save up to 70 percent more water compared to other irrigation methods. Plants love them because they provide just the right amount of water. The mechanics of an olla are simple: Bury the clay pot in the garden so that the rim is level with the top of the soil. Then fill the olla with water and allow its porous walls to gradually dissipate water into the surrounding soil. Because the clay pot's pores are small, the water does not seep from its walls freely. Instead, it moves outward by soil moisture tension. What that means is that when the soil is dry, the water is pulled out of the pot toward the soil; but when the soil is wet, water remains inside the olla. Plants are neither overwatered nor underwatered and instead get exactly what they need.

HOMEMADE OLLA

Although you can purchase ready-built ollas online, you can also make them at home by upcycling old terra-cotta pots. To prepare the pot, you'll need to seal the drainage hole with a piece of nontoxic mounting putty to prevent water from leaking from the pot. You'll also need a lid, which you can make from a terra-cotta pot saucer. If you want to use your olla to collect rainwater, remove the lid when it rains and put it back on top of the olla when the rain stops.

weeding

Controlling weeds is one of the greatest challenges for gardeners, especially if they're trying to garden organically. Weeds compete with crops for resources such as water, nutrients, sun, and space. They also provide a haven for pests and diseases. Fortunately, there are several techniques you can use to prevent and control weeds without using harmful chemicals or destructive cultivation practices.

HAND-PULL. If you have raised beds or containers, this good old-fashioned technique may be the only one you need, especially if you plant densely and leave little room for weeds to grow. When weeding by hand, your primary goal is to disturb the soil as little as possible. A lot of seeds lie dormant below the soil surface; if you churn the soil too deeply (for example, with a hoe), you risk bringing the weed seeds to the surface and giving them an opportunity to grow.

Try to remove weeds when they're small and hardly established. Larger weeds have more extensive root systems and will eventually go to seed and create more weeds. The saying "one year's seeding is seven years' weeding" is a good reminder to act preventatively and remove weeds as soon as you see them. If you're having trouble removing the entire plant by hand, try watering the soil first or using a hand trowel to dig up its roots.

GERMINATE WEED SEEDS BEFORE PLANTING CROPS. If your garden has a history of being weedy, try germinating weed seeds at the beginning of the growing season before planting your crops. About a week or two before planting, water the soil and keep it moist—just as you would if you were watering seeds. Within a few days, weeds should start to sprout; you can then remove them by hand or with a tool. If you use a tool, use one designed for cutting just below the soil surface, such as a circle hoe or hula hoe, to prevent stirring up seeds deeper in the soil. Allow the weeds to dry during the day, then toss them out in the evening. Compost leaves only; flowers, seed heads, and the roots of perennials in your compost will reintroduce weeds to your garden.

MULCH YOUR SOIL. Mulching serves many purposes in the garden. When it comes to weeding, it helps deprive weed seeds of the air and light they need to grow. Any kind of mulch will do, although if you're growing organically, it's always smart to use one that will add organic matter to the soil. Organic mulches such as compost, aged animal manure, mushroom compost, vermicompost, grass clippings, shredded leaves, pine needles, or straw will do, as will processed mulches, like cardboard or newspaper. To apply mulch, spread a 2- to 4-inch layer over the soil, keeping it a few inches away from plants' stems and trunks to prevent moisture buildup and to deter insects, slugs, rodents, and disease.

PLANT COVER CROPS. Another way to prevent weeds is to plant cover crops in the autumn. These crops are grown specifically to protect the soil, but they also serve other purposes, including adding organic matter to the soil, loosening compacted soil, balancing nutrient levels, controlling erosion, and attracting pollinators. Cover crops prevent weeds by covering open ground that would otherwise tempt weeds to take root. Common cover crops include legumes such as vetch, clover, beans, and peas. If you plant beans, they'll prevent weeds *and* enrich the soil. Cover crops can also be removed without digging, which is ideal for soil.

SOLARIZE THE WEEDS. One of the more involved ways to minimize weeds is to solarize them, or use heat from the sun to cook and kill weed seeds before they sprout. Start by dampening the soil, then spread a sheet of heavy-gauge clear plastic over it, anchored on all corners by heavy stones. Leave the sheet in place for six to eight weeks (less if you live in a warm climate), then remove it and plant your garden. Although I don't love plastic, you should be able to find something to repurpose for relatively cheap. If you do buy something new, remember to save it and reuse it as many times as possible.

feeding

Plants need nutrients to grow. If your beds are filled with a rich soil mix, they may contain enough nutrients to keep plants healthy and happy for the entire growing season. Keep an eye on your crops for signs of depletion. If your plants' growth starts to slow or their foliage changes in color, there's a good chance they need feeding. For an organic fertilizer, compost is always a good choice—if you have enough. If you don't, nourish your plants with a nutrient-rich fertilizer. Because container gardens are filled with soilless mixes void of nutrients, they will always need to be fertilized for plants to grow.

N-P-K (AND OTHER IMPORTANT ELEMENTS)

The three elements that are most vital to plants and form the basis of organic (or synthetic) fertilizers are nitrogen, phosphorus, and potassium, also known as N-P-K. Commercial fertilizers are usually labeled with three numbers, for example 3-1-2, which represents the percentages of N-P-K in each bag and in that order. A bag of 3-1-2 fertilizer, for example, contains 3 percent nitrogen, 1 percent phosphorus, and 2 percent potassium.

Knowing the ratios of those three numbers can be very important to a gardener. For example, if your plants are growing lush green leaves with no flower blooms, your soil may have too much nitrogen, in which case you would want a fertilizer low in nitrogen. If, on the other hand, your leaves are turning yellow along the leaf veins, there's a good chance your soil has a potassium deficiency, which would make it a candidate for a fertilizer high in potassium, labeled something like 0-2-5.

It helps to know a little bit about each nutrient so you can observe your plants, understand their needs, and resolve any deficiencies.

NITROGEN (N) promotes leaf growth and is responsible for making plants greener. Plants use a lot of nitrogen during the growing season, so it's usually the most depleted nutrient in the soil. If you add too much nitrogen, plants will experience vigorous growth but delayed flowers and fruit. If your plants need nitrogen, look for a fertilizer with a high first number. The higher the number, the more nitrogen the fertilizer provides. Natural sources of nitrogen include alfalfa meal, blood meal, and fish emulsion.

PHOSPHORUS (P) promotes strong root growth, which helps anchor and support plants. It also increases stem, fruit, and seed production while providing disease resistance. Plants deficient in phosphorus are often stunted, with dark green foliage, reddish purple stems or leaves, and fruits that drop early. Natural sources of phosphorus include bat guano, bonemeal, and rock phosphate.

POTASSIUM (K) supports robust growth, drought tolerance, water retention, disease resistance, enhanced flavor, and healthy development. It also enables plants to withstand extreme temperatures. Signs of potassium deficiency include yellow areas along the leaf veins and edges, crinkled and rolled-up leaves, stunted leaves and fruit, and increased sensitivity to drought. Natural sources of potassium include greensand, kelp meal, and wood ash.

TAKE A TEST

Although you can tell a lot about your soil by observing your plants, the only way to truly determine its nutrient levels is to test it. Tests are especially important if you plan to use fertilizer. Too much fertilizer can be just as damaging as too little, and using the wrong type can cause problems. Soil-testing kits can be purchased online or at most garden centers.

An earthworm test is a cheap and quick way to check the tilth, or richness, of your soil. The best time to conduct one is in the spring, when the surface of the soil is moist and temperatures have reached 50°F. Dig up about one cubic foot of soil, place it on a tray, break it apart, and check for worms. If you find more than ten worms, your soil is likely very healthy. If you find fewer than ten worms, it probably needs more organic matter. This is obviously a very low-tech test, but it's a quick way to get a read on whether your soil supports life.

ORGANIC VERSUS SYNTHETIC FERTILIZERS

Should your plants need supplemental nutrients, there are essentially two types of fertilizers you can use—organic or synthetic. Organic fertilizers release nutrients into the soil slowly, providing a steady source of nutrients that are unlikely to leach away and cause stream or groundwater pollution. They come from plants, animals, or minerals and contain a broad range of nutrients that enhance the soil ecosystem. As they break down, organic fertilizers improve the soil structure, including its ability to retain water and nutrients. Organic fertilizers are renewable, sustainable, biodegradable, and environmentally friendly.

Synthetic fertilizers are made in labs and have specific formulations that typically contain only three basic nutrients: nitrogen, phosphorous, and potassium. Because they're water soluble, they're more likely to leach into streams and waterways, and they do little to enhance the health, texture, or fertility of the soil. They release nutrients quickly, which might sound like a good thing, but some would argue they're released too quickly, creating puny plants with impressive top growth but weak roots. Synthetic fertilizers contain high concentrations of salts, which can destroy soil structure and harm soil organisms. Finally, repeated application of synthetic fertilizers can cause buildup of toxic chemicals such as arsenic, cadmium, and uranium, which are often present in small doses in synthetic fertilizers. These toxic chemicals can make their way into fruits and vegetables and ultimately your body.

I would always recommend organic fertilizers. Even though they work more slowly than synthetic ones, they're better for your garden, better for your health, and better for the planet!

Another choice when it comes to fertilizer is between dry or liquid types.

DRY FERTILIZERS are solids that need to be worked or watered into the soil to ensure they leach down toward the roots where plants can access them. They release nutrients slowly over an extended period of time. To apply them, either sprinkle them by hand or broadcast them with a lawn spreader over the soil and around plants. During the growing season, you can also side-dress them into the top 1 inch of the soil around crops. Dry fertilizers are the most commonly used and economical option, especially for small gardens and farms. Organic examples include compost, manure, bat guano, blood meal, alfalfa meal, and rock phosphate.

LIQUID FERTILIZERS give plants a quick nutrient boost and are ideal for container gardens and fast-growing veggies. You can spray them on the soil for plants to absorb through their roots, or on leaves for plants to absorb through their leaf pores. The latter way is called foliar feeding and is particularly useful for veggies during the growing season. With foliar feeding, nutrients are immediately taken up by plants, so you see quick results. Common organic liquid fertilizers include liquid fish emulsion, liquid kelp, and extracted compost teas (see page 122 to make your own compost tea).

When using commercial fertilizers, be sure to follow all label instructions for mixing rates and application. Even organic fertilizers can be overapplied, leading to fertilizer "burn," pH imbalances, and/or nutrient deficiencies.

TYPES OF ORGANIC FERTILIZERS

There are a surprising number of organic fertilizers available to gardeners, but with so many choices there's likely just as much confusion. See the chart on page 120 for some of the most popular options, with their approximate N-P-K ratios (these numbers are approximate because there's always variability in nature). When purchasing agricultural byproducts, such as alfalfa meal, corn gluten meal, or soybean meal, be sure to source organic varieties to ensure they haven't been treated with pesticides.

FERTILIZER TEA

If you're low on compost, have a tight budget, or prefer to DIY, you can make fertilizer tea using plants growing in your own garden. Comfrey tea is one of the most popular choices because comfrey is a robust perennial that grows easily and quickly. It also contains high levels of nitrogen and potassium.

To make comfrey tea, harvest enough comfrey leaves to fill a five-gallon bucket. Be sure to wear gloves, long sleeves, long pants, and shoes, as contact with the plant can irritate the skin. Weight the leaves down with a brick or large rock, fill the bucket with water, and cover it with a lid. Let the tea steep for three to six weeks, then strain the liquid and dilute it to fifteen parts water to one part comfrey tea. When you're ready to use the diluted tea, transfer it to a watering can and water your plants at their base. The tea should last for up to four weeks when stored in an airtight container.

MANURE

After compost and compost tea, manure is my fertilizer of choice. It is one of the best and most affordable options, and it can be acquired easily from local farms or stables. Avoid fresh manure, as it contains ammonia and can burn your plants. Instead, use aged manure that is dry, crumbly, and odorless. It is best to add manure to your garden in the autumn so it has time to decompose before spring planting. Sheep, horse, cow, pig, and chicken manure are all suitable; chicken manure is highest in nitrogen.

MATERIAL	N	P	K	COMMENT
Alfalfa meal	3	1	2	Adds macronutrients and trace minerals to soils
Bat guano	10	3	1	Fast uptake, great for midsummer crop rotations
Blood meal	12	0	0	Excellent source of nitrogen for fast-growing plants
Bonemeal	3	15	0	Ideal for soils that need serious amending
Chicken manure	5	2	1	High in nitrogen; be sure to compost it first
Comfrey	3	1	5	High in nitrogen, phosphorous, potassium, and calcium
Compost	1	1	1	Enhances soil life; helps make nutrients available
Corn gluten meal	9	1	1	Great soil stabilizer for winter
Cow manure	2	1	1	Nutritionally comparable to good compost; compost first
Earthworm castings	1	1	1	Contains beneficial bacteria and trace minerals
Feather meal	12	0	0	Great nitrogen booster; used pre-season to prepare soil
Fish emulsion	9	0	0	Provides nutrients quickly when leafy plants need a boost
Fish meal	10	5	4	Provides a balanced boost of minerals
Greensand	0	2	5	Great (but smelly) source of vitamins and minerals
Horse manure	4	1	1	Excellent source of nitrogen; compost first
Molasses	1	0	5	Provides nutrients for microorganisms
Pig manure	1	1	0.5	Good source of microorganisms; compost first
Rabbit manure	2	2	0.6	Excellent source of natural nutrients; compost first
Rock phosphate	0	18	0	Fantastic source of phosphorous
Seaweed	1	0	1	Good source of micronutrients; immediately available
Soybean meal	3	0.5	2.5	Great for soil maintenance

how to make compost tea

Compost tea is a natural liquid fertilizer made by soaking finished compost in water. The brewing process stimulates microbial growth and creates a nutrient-rich tea that can be used to feed plants. You can purchase compost tea at garden centers or online, or you can make it yourself using compost from your own garden. Compost tea is relatively easy, cheap, and fun to make, and it's a great activity to do with kids.

There are several ways to make compost tea. This method uses molasses and an air pump to feed the beneficial microbes in compost. If the thought of making something with an air pump sounds daunting, rest assured that all you need is an aquarium bubbler—nothing high tech or fancy! Aside from that, the supplies are simple: Just grab a bucket, some unchlorinated water, a handful of compost, a stocking, and some molasses. If your tap water is chlorinated, you will need to fill a bucket and leave it out for at least 24 hours to allow the chlorine to evaporate. Otherwise, you can use water from a rain barrel or a well. Chlorine will kill the beneficial microbes, so this step is essential!

MATERIALS

soap
5-gallon bucket
aquarium pump
about 5 gallons unchlorinated water
4 cups compost
nylon stocking
½ cup unsulfured molasses

DIRECTIONS

Use soap and water to clean and disinfect the bucket and aquarium pump. Fill the bucket with unchlorinated water. Place the aquarium pump at the bottom of the bucket and turn it on. Stuff the compost inside the nylon stocking, tie the stocking closed, and suspend it in the bucket. Add the molasses to the bucket of water. Allow the mixture to bubble for 24 hours. When the tea is ready, remove the stocking and transfer the liquid to a watering can. Use the finished compost tea to water the soil at the base of your plants.

supporting

Vegetables that climb, vine, or sprawl need vertical support to grow. There are dozens of types of supports, but since some vegetables grow better on one type than another, it's helpful to learn your options as well as how to select the right type of support for each plant. Some plants can find their way around a thick post, while others need something thinner for their short little tendrils. Some plants climb and cling to supports with sticky growth or tendrils; others need to be tied on with soft twine, fabric tape, or strips of cloth.

When tying plants to supports, remember to be gentle. A tie can rub or cut a plant's stem, especially as the plant grows or moves with wind or rain. Tie plants with a figure-eight knot—loop the tie around the plant stem, then cross the two ends and loop them around the stake before tying a double knot. The figure-eight should be taut enough to hold up the plant but loose enough to allow the plant to move as it grows.

The best time to set up supports is early in the growing season, before your plants need them. It's much easier to place a cage over a young tomato seedling than a tall, sprawling beast of a plant. There's also a greater risk of damaging branches when plants are big and need to be manipulated to fit inside cages. If you set up a cage early on, it will slide right over the young plant, which can then grow neatly within and up its walls.

Always err on the side of providing too much support. Flimsy structures can collapse under the weight of heavy plants. Taller supports also support higher yields by allowing plants to grow to their full potential. When it comes to the size of supports, the only limiting factor is you! You don't want to have to drag out a ladder to harvest your crops, so design your supports with ease in mind.

TYPES OF SUPPORT AND PLANTS THAT NEED THEM

A few of the most common types of supports are stakes, cages, teepees, trellises, and arbors. You can buy them, or you can make them yourself from materials like wood, bamboo, or metal, or upcycled ladders, bed frames, old screen doors, windows, cattle panels, or untreated pallets. Some of the most common veggies that need support are cucumbers, melons, peas, pole beans, squash, and tomatoes.

CUCUMBERS. The best way to support cucumbers is to train them up a trellis on the edge of a bed or along a wall. Cucumbers are heavy and sprawling, so the ideal trellis should be about 6 feet tall and strong. Although cucumbers will naturally climb, you may need to help them by weaving their vines through a frame or attaching them to structures with soft plant ties.

MELONS. Melons grow on sprawling vines that can overtake a garden if you let them. The logical solution is to grow them vertically on a support. Melons grow well on a strong, sturdy trellis. But as the fruits mature and get heavy, they become susceptible to falling and cracking when they hit the ground. To prevent this from happening, gardeners create slings to hold them in place. The most common way to make a melon sling is to use lengths of stretchy fabric, like nylon pantyhose or cut-up cotton T-shirts, to cradle the fruits and tie them to a trellis. Larger melons like muskmelons and watermelons are extremely heavy when mature and are best left to grow on the ground.

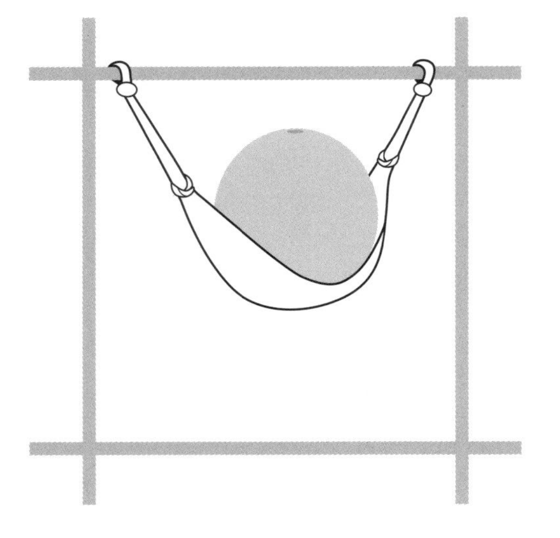

PEAS. Vining peas like shelling peas, sugar snap peas, and snow peas climb with the help of slender tendrils that wrap around a support and pull the plant upward. It's important to support peas to direct the growth of the pea vine, keep it off the ground, and make it easy to pick the peas. The cheapest and easiest way to support peas is by using materials you already own—chain-link fences and chicken wire are excellent options. You can also place stakes, such as sturdy sticks, bamboo poles, steel rebar, or spare timber, in the ground a few feet apart behind your peas.

SQUASH. Trellising is a great way to support squash plants. A popular option for a squash trellis is welded wire fencing over an A-frame or suspended from the side of a fence or building. Like cucumbers and melons, squash plants grow tendrils that will grab hold of and climb anything nearby. Mature winter squash grown vertically will also need a sling to keep it from falling to the ground (very heavy squash such as pumpkins should simply be allowed to spread on the ground). Summer squashes and smaller winter squashes, such as acorn squash, will not need quite as much reinforcement and they won't need slings. A tomato cage is often all you need to keep these squashes from taking up a lot of space in the garden.

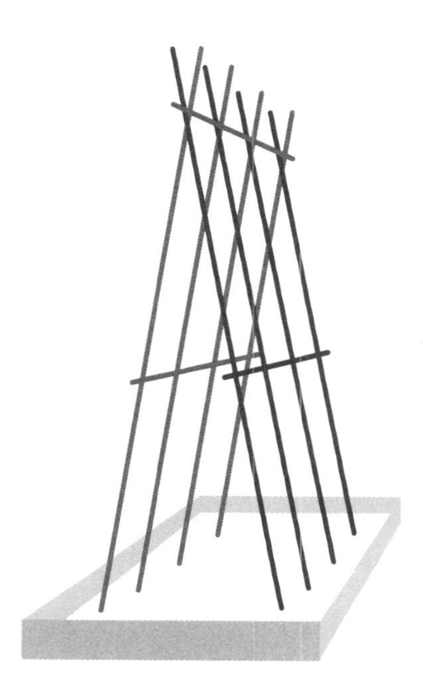

POLE BEANS. Unlike bush beans, pole beans require special support for their climbing vines. Stakes are the traditional support for pole beans. They don't need to be fancy. Just purchase stakes—or cut long, slender pieces of wood—6 to 8 feet long. Hammer them into the ground where you intend to plant the beans, then plant the seeds at the foot of the stake. The beans will wind around and around the poles without needing much help from you. A fun way to grow beans up poles is in the shape of a teepee, like the one on page 199. A trellis also makes a good climbing structure for pole beans. One of my favorite types of trellises for beans is a crisscross trellis, where you cross pairs of 6- to 8-foot-long bamboo stakes at the top to form a triangle, space each triangle 6 to 9 inches apart, and join them across the top with another bamboo stake. Once complete, it looks like a series of *A*s connected by a crossbar. As with all bean supports, it's best to plant one or two seeds at the base of each stake.

TOMATOES. Most people don't realize it, but tomatoes are vines with no means of attaching themselves to supports. Left to their own devices, they'll grow into a tangled mess, which might explain why gardeners have come up with so many ways to support them. Tomato plants typically fall into one of two categories: determinate or indeterminate. Determinate tomatoes grow short and busy, whereas indeterminate tomatoes grow large and tall. With determinate varieties, you can often get away with using a single stake. A tomato cage also works well, the most common being a cone-shaped, heavy-gauge wire "tomato basket." For indeterminate varieties, it's best to use something sturdier like a vertical trellis; a double-decker cage that stands about 8 feet tall will work fine too. If you use a trellis, you'll need to tie or weave the plant through it since it doesn't have tendrils. With cages, simply tuck the branches through the openings and let the wires support them.

how to build a simple trellis

There's no need to go out and buy a fancy support structure when you can easily make one yourself. A simple trellis can be made with a few pieces of lumber, a bit of netting, and an extra pair of hands. Most trellis netting is made from polyester, but you can also find jute netting if you would prefer a more sustainable option. This particular trellis is sturdy enough to support climbing flowers like morning glories and climbing nasturtiums as well as vining vegetables such as peas, cucumbers, or pole beans. For a sturdier support, I would recommend using thicker 2x2 lumber for supports and a piece of cattle panel (attached to the trellis with galvanized wire) instead of netting. Cattle panel can be purchased at any farm supply store. These instructions are for a 5-by-7-foot trellis, but you can adjust the dimensions as desired to fit your space.

MATERIALS AND TOOLS

shovel or post hole digger
3, 8-foot pieces of 1x1 untreated lumber
jute twine or floral wire
5-by-7-foot piece of trellis netting
scissors or wire cutters
drill or screwdriver
deck screws

DIRECTIONS

1. Choose a location for your trellis. Dig two 3-foot-deep post holes 7 feet apart.

2. Place a 1x1 in each hole and backfill the holes tightly with soil.

3. Using twine or wire, secure the long side of the netting to the third 1x1, which will become the top support beam.

4. Lift the top beam onto the two upright posts and screw it to each post with deck screws.

5. Attach the sides of the netting to the two side posts with twine or wire.

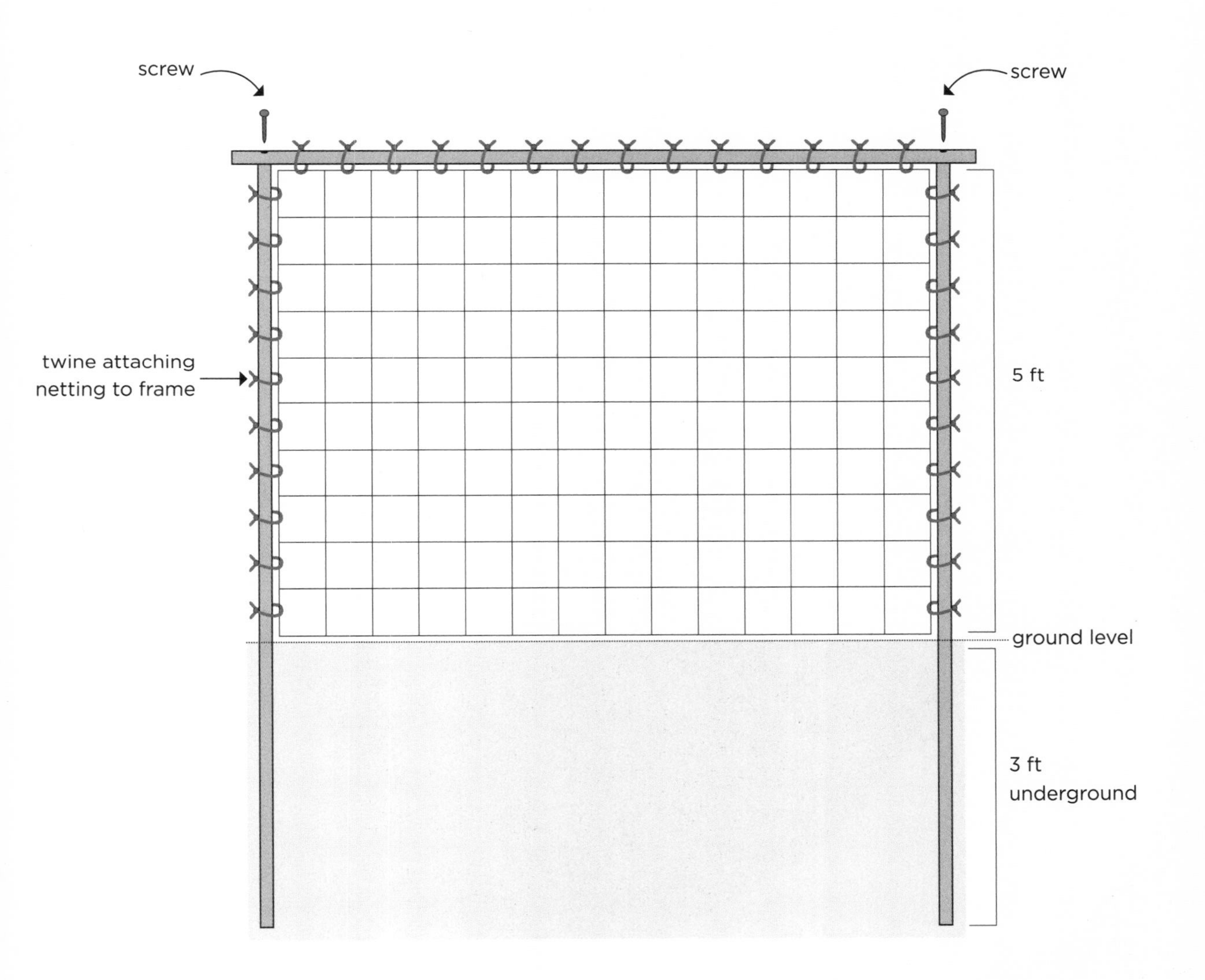
screw
screw
twine attaching netting to frame
5 ft
ground level
3 ft underground

pruning

Pruning plants is one of the most difficult tasks for gardeners, especially because it can feel like you're discarding—or potentially hurting—the plants you worked so hard to grow. But thinning crowded seedlings, pinching back new growth, and removing excess foliage can lead to healthier, more productive plants. In addition to removing dead, diseased, and damaged foliage, pruning improves air circulation, makes it easier to spot and deal with pests and disease, allows plants to put their energy into new, healthy growth, and prevents the garden from becoming a scraggly, unruly mess.

Pruning techniques vary from crop to crop, but a few best practices apply to them all:

- **DISINFECT PRUNERS** between cutting different plants and especially after cutting diseased-looking parts. Dip the blades in a solution of hydrogen peroxide (diluted with water at a ratio of 1:10), then rinse them with water.
- **REMOVE NO MORE THAN ONE-THIRD OF A PLANT** at a time to ensure it has enough foliage for photosynthesis, which it needs to do to continue to grow.
- **REMOVE OLDER, BOTTOM LEAVES** as they naturally turn yellow and brittle, both to prevent disease and to redirect the plant's energy toward new growth.
- **LEAVE FRUITING BRANCHES ALONE** and instead prune branches that are mostly leaves.

If you feel a little intimidated by the prospect of having to prune your plants, rest assured that there's not much you need to do in a vegetable or herb garden. Most vegetables don't need to be pruned at all, and others just need to be kept neat and tidy with some thinning, pinching, and trimming. Serious pruning is mostly relegated to shrubs and trees. Here are a few of the most common pruning tasks for a vegetable and herb garden.

THINNING SEEDLINGS

The first stage of pruning happens early in the season after sowing seeds. Once the seedlings emerge, thinning is usually a must—otherwise, plants will compete for nutrients and space as they grow. To thin seedlings, look for the weakest-looking sprouts and either gently pinch them off at the base of the plant or use a pair of scissors to cut them out at soil level.

PINCHING HERBS

Like all plants, herbs want nothing more than to reproduce. Left to their own accord, they'll cease growing, produce flowers, and go straight to seed. That's because the fast-growing tips send chemical signals to secondary buds, telling them to stop growing. In nature, this adaptation makes them supremely fit for survival, but in the garden, it makes them kind of obnoxious! If they go to seed too quickly, you're left with tall, lean plants that have few if any lateral branches, which is hardly the outcome gardeners seek.

Luckily, there's a way to keep herbs in their growth stage longer so they develop into fuller, more productive plants. It's called pinching back (or tipping) and involves using your fingers to remove the topmost set of leaves. Pinching encourages plants to continue growing and prompts them to branch at the pinch point, resulting in fuller plants with more leaves.

Harvesting achieves the same results as pinching back, but most gardeners don't start harvesting herbs until they're nice and full. Pinching back can start when herbs are 6 to 8 inches tall, or when three sets of leaves develop on the stems. Either way, both harvesting and pinching back should be done as early and as often as possible to optimize growth.

PRUNING TOMATOES

Of all the veggies in the garden, tomatoes are the ones most commonly pruned. But not all types of tomatoes need to be pruned. In general, pruning is most helpful for indeterminate types, which are large plants that continue to grow taller and produce fruit throughout the growing season. Determinate, or bush, tomatoes tend to be smaller and more manageable. They also produce tomatoes all at once. If you prune determinate types, you risk reducing the harvest.

Most tomato pruning involves removing "suckers"—the shoots that sprout from where the main stem and side branches meet. If your goal is to maximize your harvest, prune suckers sparingly. Suckers that are not removed will grow into additional stems and create a full, bushy plant. But if space is limited, or if you're supporting tomatoes with stakes, it's best to remove all the suckers and prune the plant back to one or two main stems. The main stems will grow strong and sturdy and will be easy to secure to supports. Although you will probably harvest fewer fruits from a pruned tomato plant, each fruit will be larger than if the plant had not been pruned.

DEADHEADING FLOWERS

Pruning flowers is called deadheading and involves removing faded or dead flowers from plants. As flowers shed their petals and begin to form seed heads, energy is focused into seed production. Regular deadheading brings a halt to this process and redirects a plant's energy into growing new flowers. To deadhead flowers, simply pinch or cut off the flower stem below the dead flower and above the first set of healthy leaves. The result is healthier looking plants and recurring blooms. If you prune them regularly, most annual and perennial flowers will continue to bloom throughout the growing season.

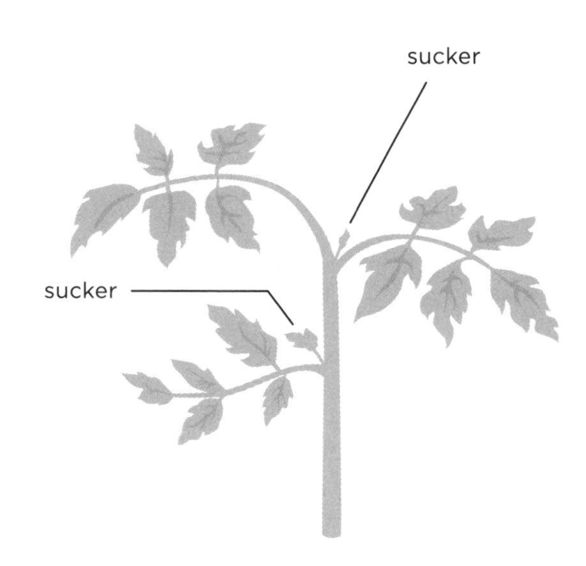

hand-pollinating

Plants in the cucurbit family, such as cucumbers, squash, melons, and pumpkins, have a reputation for being poor pollinators. If you've ever tried growing them, there's a good chance you've ended up with rambling plants, lots of flowers, and no fruit! There are several factors that contribute to poor pollination. To make sense of them, it's helpful to understand how cucurbits pollinate in the first place.

Members of the cucurbit family have separate male and female flowers that must cross-pollinate to produce fruits. Cross-pollination happens when the pollen from a male flower is transferred to a female flower. The two types of flowers are easy to tell apart—if you look under the blooms of each flower, males have a straight, slender stem, while females have a swollen base that resembles a tiny squash. Once fertilized, female flowers turn to fruit; if they don't get pollinated, they wilt and fall off the stem.

Poor cross-pollination is caused by many factors. Sometimes there just aren't enough pollinators in an area to transfer pollen from male to female flowers. Other times weather conditions can cause the premature death of pollen grains. Wet weather or badly timed watering can also affect pollen quality and quantity.

If your cucurbits have a history of being poorly pollinated or you want to ensure maximum fruit production, you can hand-pollinate the female squash flowers yourself. If you have children, this is an excellent way to involve them in gardening. My daughter Eloise is usually in charge of this task, and let me tell you, she takes it very seriously!

There are several methods for transferring pollen from one flower to another. What they all have in common is that the pollen is moved from the male flower to the female flower, and not the other way around. The best time to transfer pollen is during daylight hours, when flowers are open. Use something soft, like a cotton swab or a small paintbrush. Locate an open male bloom, gently brush your tool against its stamen to coat the tool with pollen, then brush the pollen against the stigma of a female flower. To maximize yields, repeat this step for every female flower on your squash plants. If you've succeeded in pollinating your plants, the female flower will wilt and fall off, but the squash will remain on the stem.

preventing and managing pests

Nothing bugs a gardener more than pests! The heartbreak of finding holes in your cabbage or aphids on your corn is demoralizing, to say the least, especially after putting weeks, even months, of your heart and soul into planning, planting, and tending your crops. And though you might be tempted to reach for something poisonous to get rid of the pests, pesticides aren't a sustainable solution. They're not just toxic to pests; they're also dangerous for pets, wildlife, and humans. Pesticides contaminate water and disrupt healthy, natural ecosystems in and around your garden, to say nothing of how they destroy the delicate and vital soil food web that sustains your plants.

Organic gardeners take a different approach to pest management. Instead of focusing on killing pests, first we look for gentle ways to prevent them, and then we employ nontoxic control methods to manage them. Here are a few of the most common (and gentle) ways to reduce damage from pests.

PLANT PEST-RESISTANT CROPS. Gardeners often overlook one of the most important weapons for preventing pests, which is carefully selecting pest-resistant varieties of plants. If squash borers are a problem in your area, taking a few extra minutes to research squash varieties that show resistance or tolerance to this pest can save you hours of grind *and* increase your garden's yield.

GROW HEALTHY PLANTS. One of the best ways to prevent pests is to meet your plants' needs for healthy soil, sufficient sunshine, and adequate water. When a plant's needs aren't met, it becomes stressed and emits a chemical signal that pests pick up on. Although pests will also infest healthy plants, they tend to target weaker plants that are less likely to survive an attack.

INTERPLANT CROPS. Pests are often plant specific. They also enjoy monocrops (a single crop grown year after year in the same area), which is why industrial farms are often heavily sprayed with pesticides. Instead of monocrops, alternate rows of vegetables with rows of beneficial-insect-attracting and pest-repelling herbs and flowers (see "Companion Plant," at right). If you plant all of your cabbage in one bed, it's much easier for cabbage moths to zero in on them and spread like wildfire.

ROTATE CROPS. Moving crops to different locations from one season to the next can reduce pest populations in the garden. At the end of a growing season, many pests will leave their eggs in the soil near their host plants. That way, their offspring will have their favorite foods nearby when they hatch in the spring. If the crops they feed on are easily available, pest populations will grow quickly. But if you plant something unexpected and unfavorable in that spot instead, pests may starve to death before they find their food source.

KEEP FOLIAGE DRY. Wet leaves invite insects and fungal infestation. Remember to water the soil early in the day and at the base of your plants to ensure foliage stays dry at night.

DON'T OVERFERTILIZE. Be careful not to apply too much fertilizer, as excess nutrients can attract certain pests. Too much nitrogen, for example, results in rapid growth and succulent plant tissues, which are especially favorable to aphids and spider mites. An imbalance of phosphorus encourages egg reproduction in spider mites. The best way to prevent overfertilizing your plants is to amend the soil with organic matter and use organic fertilizers rather than fast-acting synthetic ones.

COMPANION PLANT. Employ this age-old tradition of placing specific plants near each other to attract beneficial insects and repel pests. For example, plant marigolds with tomatoes to deter nematodes, slugs, and tomato worms. Plant nasturtiums to draw aphids away from cucumbers, tomatoes, squash, and courgettes. Planting alliums (onions, chives, garlic, or shallots) near tomatoes helps deter aphids, weevils, and moles. Planting strong-scented herbs is also an excellent way to keep pests at bay.

ATTRACT BENEFICIAL ORGANISMS. Perhaps the most natural way to prevent pests from ravaging your garden is to create an environment that attracts their predators. Birds, bats, frogs, garden snakes, spiders, and a host of soil organisms and insects can all help keep pest populations in check. Below I'll introduce you to some beneficial organisms and share ways to lure them into your garden.

welcoming beneficial insects

Bugs get a bad rap. In the garden, we call them pests, curse them when they eat our vegetables, and even go as far as to poison them with chemical pesticides. But according to researchers, 97 percent of all bugs are beneficial in some way—they either do no harm, provide food for other animals such as birds, or prey upon the insects that destroy gardeners' crops. Many insects are pollinators, and when they disappear, our crop yields suffer. When we use chemical insecticides to get rid of pests, we end up eliminating the bad guys (for now), but we also eradicate the good guys that serve as pollinators and natural pest control in the garden.

Here's a list of some of the best beneficial insects, with tips for how to attract them.

ASSASSIN BUGS. Assassin bugs look like a strange mix between a praying mantis and a squash bug. They use their sharp mouthparts to prey upon many different types of insect pests in the garden, including leafhoppers, aphids, larvae, boll weevils, and insect eggs. As beneficial as they are in the garden, assassin bugs will bite if handled or disturbed. Their bite is not considered dangerous, but it can be painful. Assassin bugs love Queen Anne's lace, daisies, and alfalfa.

BRACONID WASPS. These parasitic wasps lay their eggs on the backs of tomato hornworms and other caterpillars. If you see a parasitized caterpillar with little white spots on its back, don't kill it. Wasp larvae will take care of it for you. If you don't have parasitic wasps in your garden already, you can purchase them commercially. Plant dill, fennel, and other parsley family plants to keep them around.

GROUND BEETLES. Many beetles live on the ground or in the soil, where both adults and larvae feed on a wide range of pests, including aphids, caterpillars, fruit flies, mites, silverfish, slugs, nematodes, and weevils. The best way to attract them is to avoid using herbicides and insecticides in or near your garden.

HONEYBEES. As pollinators, honeybees are among the most well-known beneficial insects. Adults have gold and black bodies with translucent wings. You can attract them to your garden by planting flowers and herbs rich in nectar and pollen, such as asters, cosmos, calendula, zinnias, sunflowers, mints, lavenders, borage, rosemary, echinacea, and sages. Bees are especially drawn to blue, purple, and yellow flowers. A water source in dry weather is helpful too.

HOVERFLIES. The hoverfly looks like a tiny wasp without a stinger. It gets its name from the adult's habit of hovering around flowers. Adults feed on pollen and nectar and are important pollinators. Their larvae are voracious predators and feed on aphids, beetles, caterpillars, sawflies, and thrips by sucking the juice from their victims. Hoverflies aren't all that hard to attract. Just grow some flowers and they will come!

LACEWINGS. Adult lacewings feed on pollen and nectar. Their larvae prey upon soft-bodied garden pests like caterpillars and aphids. They are attracted to composite flowers, such as asters, black-eyed Susans, goldenrod, and yarrow.

LADYBIRDS. Despite their dainty name, ladybirds are merciless predators in the garden. A single adult ladybird will eat fifty aphids a day; her younglings will eat thousands of aphids, spider mites, and whiteflies in the same period of time. They can be lured to the garden by food in the form of pollen and pests. Although it might seem counterproductive, leaving aphids and other pests alone in the garden will help attract ladybirds. Ladybirds like the blooms of many plants, including chives, coriander, cosmos, dill, fennel, marigolds, and yarrow.

PRAYING MANTISES. A praying mantis will prey on a host of insect pests, including aphids, beetles, caterpillars, earwigs, leafhoppers, flies, and moths. The only downside to praying mantises is that they don't differentiate between good guys and bad guys and will also eat beneficial creatures such as butterflies, bees, and hummingbirds—and even their own kind! Since they prey on other insects, the best way to attract praying mantises to your garden is by creating an organic, bug-friendly garden. They're also enticed by plants in the rose and raspberry families.

SOLDIER BEETLES. Soldier beetles are an important predator of Mexican bean beetles, caterpillars, cucumber beetles, grasshopper eggs, and aphids. Like many beneficial bugs, they are attracted to plants that have compound blossoms. Also, because they spend the majority of their life cycle in the soil, the best way to keep their numbers up in the garden is to take good care of your soil. Practicing organic gardening and using no-till methods will go a long way in protecting these beetles.

SPIDERS. Though they're technically arachnids, not insects, spiders are very adept at pest control. Jumping spiders and wolf spiders are especially good at keeping pest populations under control. One way to attract spiders to your garden is to create a habitat that protects them from the elements. Adding a loose layer of mulch or grass clippings between plants, for example, provides them shelter from dry air, direct sunlight, and heavy rains.

TRICHOGRAMMA WASPS. These tiny parasitic wasps lay their eggs inside the eggs of more than two hundred different species of moths. The developing larva then eats the host from inside its egg. Purchase them at garden centers or online, then release them when their host species are most likely to lay their eggs, usually at the beginning of spring.

attracting beneficial organisms

Insects aren't the only beneficial organisms in the garden. If you avoid using pesticides and herbicides, your garden will become a living, thriving ecosystem, full of good bacteria, microorganisms, and worms in the earth, as well as arachnids, small creatures, and birds above the soil line.

More than 90 percent of the living creatures in a healthy garden are beneficial to the plants growing there, and they work arduously to prey on the small percentage of organisms that are actually pests. Robins and thrushes eat caterpillars and snails; frogs and toads munch on slugs, woodlice, and flies.

Like all living creatures, beneficial organisms have similar basic needs. They need food in order to reproduce and feed their offspring; shelter to raise their young and hide from predators; and water to stay hydrated and healthy. By providing these things, your garden can become an inviting home for them. Here are a few tips to help you turn your garden into a diverse community of beneficial organisms.

CONSIDER THE GARDEN AN ACTIVE ECOSYSTEM. Instead of focusing on battling bad bugs with chemical pesticides, switch your focus to creating a community of living organisms that establish a natural balance in your garden.

BUILD A HOUSE IN THE SOIL. One handful of healthy soil holds billions of different living organisms hard at work, making soil healthier while preying on nonbeneficial organisms. These bacteria, fungi, protozoa, mycorrhizae, earthworms, arthropods, and nematodes need food, shelter, and water too! The best way to meet their needs is to provide them with plenty of organic matter throughout the year.

PLANT THEIR FAVORITE FOODS. Plant pollinator-friendly plants that provide nectar and pollen to beneficial bugs and birds. Focus on native plants that have evolved with pollinators in your region. Nonnative plants may not provide pollinators with enough nectar or pollen or may be inedible to butterfly or moth caterpillars.

SET OUT FOOD FOR BIRDS. Landscape in and around the garden with plants that produce seeds and berries that mature at different times of the year. When cleaning the garden for autumn or winter, leave the seed heads of annual and perennial flowers for birds to feed on during winter. Supplement naturally available food with commercial or homemade bird food to encourage birds to visit your garden year-round. Remember to keep feeders and tables clean so the birds stay healthy and disease-free.

CREATE SHELTER OR HABITAT FOR BENEFICIAL ORGANISMS. Beneficial organisms need places to feed, hide, reproduce, and overwinter. Consider creating borders of diverse vegetation with flowers, shrubs, and grasses. Build a garden pond as a home for frogs and toads. Create a wild area with piles of stones, logs, and leaves, where insects can nest and hibernate for the winter. Install wildlife homes and nesting boxes for frogs, birds, bats, and insects. Allow a dead tree stump to remain standing to provide a home for insects, rodents, and birds.

PROVIDE A WATER SOURCE. Provide a reliable source of clean water for birds, bees, and butterflies using saucers, birdbaths, fountains, or in-ground water features. Change the water often to avoid increasing mosquito populations.

TOLERATE SOME PESTS. Remember that having a few pests is actually a good thing. Beneficial insects are attracted to gardens that have their favorite food on the menu. Try to tolerate some pests, especially in early spring, when they will attract beneficial organisms that are preparing to reproduce for the year.

ADD DOMESTICATED FOWL. You can introduce chickens or ducks to your garden. They can provide nutritious eggs and manure for compost while foraging on slugs and snails.

LIMIT PESTICIDE USE. Choose nonchemical methods of insect control, including hand-removing pests, pruning away infestations, and using row covers.

pest control methods

If you can't prevent pests, use the following methods to *control* them. Try to start with the simplest and least harmful methods and then work your way to more complicated treatments, if needed. Organic control methods can be physical, biological, or chemical in nature.

BARRIERS (PHYSICAL). Physical barriers are one of the best lines of defense against pests. Fences can be extremely effective at deterring vertebrate pests such as deer, rabbits, squirrels, and chickens. Bird netting prevents winged visitors from eating fruits and tender young seedlings, and floating row covers protect young seedlings from insect attacks. Row covers are especially helpful against mobile pests such as cabbage moths, Mexican bean beetles, squash bugs, and tomato hornworms.

CLOCHES AND COLLARS (PHYSICAL). A cloche is a small transparent cover that sits over a plant, creating a shield from pests such as squirrels, rabbits, and birds. Cloches can be made from wicker, wire, plastic, or glass. Collars can be used to protect plants from cutworms or other pests that attack plants at the soil level. Collars can be made from empty toilet paper rolls or tuna cans; simply slip one around the stem of a plant and press it firmly into the soil.

BEER TRAPS (PHYSICAL). Beer is highly attractive to many soil-dwelling pests, such as woodlice, slugs, and millipedes, and can be used to trap them before they devastate your favorite plants. To make a beer trap, remove the top from a shallow can such as a tuna or cat food can, clean the can, and bury it in the soil, leaving ½ inch above the soil surface. Fill the can one-third full with beer and wait for the slugs to find it. Dispose of the beer and drowned slugs in your compost heap.

HANDPICKING (PHYSICAL). Many gardeners find handpicking pests like cabbage moth eggs and slugs to be an effective, simple, and free organic pest control method. If bugs make you squeamish, wear gloves or invest in a handheld, battery-operated bug vacuum. Once you remove the bugs, drop them in a bucket of soapy water to kill them, then discard the soapy water away from your garden.

HOSING OFF (PHYSICAL). If handpicking is too much for you, try removing small-bodied insects like aphids and spider mites by washing them off with a hose. It may take a few treatments to get rid of them completely, but it works if you stick with it.

PRUNING (PHYSICAL). Pruning is another variation of manual removal and works well when an infestation of insects is concentrated on a few leaves or branches of a plant. When pruning, take care to avoid removing too much of the plant so it can continue to photosynthesize.

STICKY TRAPS (PHYSICAL). These glue-based traps are used to attract, catch, and immobilize pests. To be effective, traps must be clean and placed every 3 to 5 feet in the garden. Yellow traps attract whiteflies, fruit flies, leaf miners, aphids, and thrips. White traps lure whiteflies, cucumber beetles, and flea beetles. If you like to DIY, you can easily make sticky traps yourself! A quick online search will give you some ideas.

PARASITIC NEMATODES (BIOLOGICAL). These beneficial organisms are effective against squash vine borers, cut worms, root maggots, and weevils. Nematodes are ingested by pests. Once inside them, the nematodes release bacteria that kills the pests within a day or two. Parasitic nematodes occur naturally in the soil, but you may need to buy more to manage an invasion.

BACILLUS THURINGIENSIS (CHEMICAL). Also known as Bt, *Bacillus thuringiensis* is a naturally occurring bacterium found in many soils. There are several different types, which are used to kill a specific insect or class of insect. Bt is applied as a liquid that you spray on plants you want to protect. To be effective, the insecticide must be ingested by the insect. The most popular strain is called Bt var and kills several hundred types of caterpillar.

DIATOMACEOUS EARTH (CHEMICAL). Diatomaceous earth is a fine, talc-like powder made up of the fossilized remains of a tiny aquatic phytoplankton called diatoms. Diatomaceous earth comes in many grades. For safety, I use the food-grade variety in my garden. Although diatomaceous earth feels smooth to humans, it creates abrasions on the exoskeleton of insects, which deteriorates their body's protective outer layer, causing the insect to desiccate and die.

INSECTICIDAL SOAPS (CHEMICAL). Insecticidal soaps work by dissolving the cuticle that holds moisture inside an insect's body. To be effective, they must be in liquid form when they come in contact with a pest—insecticidal soaps do not work once they've dried on crops. They're sold online and at garden centers and can be used to combat mites, aphids, whiteflies, and other soft-bodied insects.

You can also DIY your own soap spray. The advantage to making it yourself is that it saves money and it gives you complete control of the ingredients. (Some commercial soaps contain an excess of unrecognizable ingredients.) To create your own soap spray, combine 1 tablespoon pure castile liquid soap per 1 quart water. You can intensify the soap spray by adding 2 to 5 drops of essential oil per quart. Peppermint, lavender, orange, tea tree, rosemary, and eucalyptus are just a few of the essential oils that act as natural insect repellents.

NEEM OIL (CHEMICAL). Neem oil is a concentrated plant-based oil, extracted primarily from the seeds of the neem tree. The oil is particularly effective against aphids, beetles, caterpillars, leaf miners, thrips, spider mites, mealybugs, scale, and whiteflies. When applied directly, the oil coats insects' bodies and kills them directly or interferes with feeding, molting, mating, and egg-laying. The shiny protective coating that neem oil adds to leaves makes them less susceptible to fungal diseases, such as powdery mildew, rust, or blight. When used properly, neem oil does not kill beneficial insects.

REPELLENTS (CHEMICAL). You can keep some pests away by repelling them with oils, sprays, or scent dispensers that hang near or on crops you want to protect. Neem oil (see above) is a common repellent that works by interrupting an insect's ability to grow, develop, and lay eggs. Other types of repellents are made from the extracts of plants like garlic, marigolds, onions, rosemary, and sage.

SLUG PELLETS (CHEMICAL). Used to kill snails and slugs, slug pellets are an organic iron phosphate–based product that is safe for pets, wildlife, and people. Just sprinkle the little white pellets in problem areas. Not only will they eradicate slugs, but they also fertilize your plants as they break down in the soil. Just be careful not to overuse them, as they can result in a buildup of iron phosphate and harmful chelates in your soil.

common pests and solutions

If you notice holes in your leaves or bright-colored bugs hanging around your flowers, it's helpful to be able to identify them and know know how to manage them if they are pests. Here I've listed the most common pests I've encountered as well as the best control methods to keep them from wreaking havoc in the garden.

INSECT PESTS

APHIDS. These tiny, pear-shaped pests come in lots of different colors—green, yellow, brown, red, gray, or black. They pierce stems and suck the nutrient-rich sap from the plant, leaving behind curled or yellowed leaves, deformed flowers, or damaged fruit. They also secrete a sticky substance called honeydew that attracts ants and black mold. Aphids feed in large groups and reproduce multiple generations in one year; infestations can quickly get out of hand. Manage them by spraying them with a hose, attracting lacewings or ladybirds, using yellow sticky traps, or spraying them with insecticidal soap.

CABBAGE MOTHS. These inch-long gray moths lay eggs on broccoli, cabbage, cauliflower, and other plants in the brassica family. In their caterpillar stage, they are green with white stripes down their sides and move with a looping motion. They chew holes in leaves and, in numbers, can cause complete defoliation. Manage them by covering plants with floating row covers, hand-removing caterpillars, attracting beneficial wasps, or spraying them with Bt.

CARROT RUST FLIES. As adults, these flies are small, shiny, and black with orange heads and legs. The larvae are tiny, beige-colored maggots. Adult flies lay eggs near many vegetable crops, including carrots, celeriac, celery, parsley, and parsnips. Their larvae feed on crop roots, leaving tunnels and scarring behind. Manage them by employing crop rotation, using floating row covers, or introducing beneficial nematodes to the garden. Interplanting carrot-family crops with onions, garlic, and chives is believed to deter carrot rust flies from laying their eggs nearby.

CUCUMBER BEETLES. As adults, these beetles measure about a quarter inch long and are bright yellow with spots or stripes, depending on the species. Their larvae live underground and out of sight. As the name suggests, these beetles feed on members of the cucumber family, including cucumbers, squash, melons, pumpkins, and gourds, and create small, ragged holes in leaves and flowers. They also transmit bacterial wilt. Manage them by planting bacterial-wilt-resistant cultivars or installing floating row covers (until your plants flower; then you'll need to remove them for pollination).

CUTWORMS. These two-inch-long caterpillars emerge at night to chew through the stems of young plants at the soil line. They disturb a wide range of plants but are especially drawn to broccoli, cabbage, kale, and tomatoes. Manage them by employing crop rotation, handpicking caterpillars, spraying Bt or horticultural oil, or setting out collars.

FLEA BEETLES. These tiny black or brown beetles move quickly and hop like fleas. They affect many different types of plants and make small, round holes in plant foliage. Because they feed in large groups, they can quickly skeletonize plants. Their larvae live underground and can consume plant roots, too. Manage them by practicing crop rotation, using yellow sticky traps, or employing beneficial nematodes.

LEAF MINERS. The tiny brown or green larvae of this insect tunnel between layers of leaf tissue, creating their telltale squiggly lines on leaves. Different species of leaf miners feed on different plants, but common host plants include beetroot, blueberries, Swiss chard, nasturtium, and spinach. Manage them by removing leaves from infested plants and placing row covers over susceptible plants. Neem oil can sometimes help as a last resort.

SLUGS AND SNAILS. Although they're not insects, slugs and snails are common pests that feed on the tender leaves of many different types of plants, both at night and in wet, rainy weather. They can be gray, black, orange, brown, tan, or mottled, and often leave a slime trail behind. Snails and slugs leave irregular holes in leaf margins or centers. Prevent and manage them by watering plants in the morning so leaves will have all day to dry. Otherwise, attract birds, frogs, and toads to eat them; set out beer traps; broadcast slug pellets; or hand-remove individuals and drop them in a bucket of soapy water.

SQUASH BUGS. These brownish-gray or black bugs suck the juices from the leaves and stems of cucumbers, melons, pumpkins, squash, and courgettes until they wilt, dry, and turn black. Prevent and manage squash bugs by planting resistant varieties, rotating crops, using trellises to keep growing vines off the ground, removing egg clusters from leaves, and resisting the urge to mulch around plants, as mulch gives squash bugs a place to hide and overwinter. It also helps to use row covers until plants start flowering.

TOMATO HORNWORMS. Adults are large, nocturnal moths with brown or gray wings. Their larvae are bright green caterpillars that can grow up to four inches long. Hornworms prey on members of the tomato family, including aubergines, peppers, potatoes, and tomatoes. Signs of this pest include eaten leaves, often toward the top of plants. Manage them by handpicking and dropping them in soapy water (or feeding them to your chickens) or spraying leaves with Bt.

NON-INSECT PESTS

CROWS. As omnivores and opportunists, crows will eat anything, including seeds, fruits, insects, grain, berries, earthworms, and even young birds. If your seedlings go missing, especially corn seedlings, there's a good chance you have crows. You can protect fruit crops with 4-inch mesh bird netting or use visual scare tactics to keep crows away. Shiny pie tins, CDs hung from trees, scarecrows, Mylar balloons, and flags have all been shown to startle crows. Protect young seedlings with floating row covers, and protect ripening corn by placing a bag over each ear after the silk turns brown.

DEER are one of the most destructive garden pests. When they're not eating your crops, they're trampling them. Deer visits can be identified by hoofprints, piles of deer droppings, and leaves with ragged bite marks. The best way to keep deer out of your garden is to build a tall, sturdy fence, up to 8 feet high. Spray repellents also work. For a homemade recipe, mix three raw eggs in a gallon of water and spray it on your plants.

MOLES. These ground-dwelling carnivores feed on insects, grubs, and soil organisms rather than crops, but they can ruin a garden with their underground tunnels, which provide access for other rodents. You'll know you have a mole problem if you spot surface tunnels or ridges near or in your garden. If you have a lot of moles, it can be a warning sign that your soil is highly populated with pests. The only surefire way to prevent moles is by using traps, preferably humane ones. If you don't want them coming back, release them at least five miles away. Letting your cat wander the garden will work too, although it won't be nearly as humane!

RABBITS are voracious little nibblers who love to graze low to the ground on young tender growth. Rabbit damage can be identified by a clean, angled cut on the ends of leaves and twigs. The best way to keep rabbits out of the garden is to build a fence that is at least 4 feet high and 6 inches below the ground (since they burrow). Bend the top foot of the fence away from the garden like a security fence to keep them from climbing and jumping over it.

SQUIRRELS. These relentless critters can drive a gardener nuts, especially because there's not much squirrels won't ravage in the garden. They'll eat fruits, vegetables, flowers, bulbs, even insects! If your fruits go missing, your bulbs get dug up, or your crops look damaged, a squirrel is the likely culprit. To protect your veggies from squirrels you can install a wire fence, buried about a foot into the ground. Some gardeners use chew-proof netting or row covers to keep squirrels from eating their crops. As far as repellents go, natural commercial varieties are available, usually made from the urine of squirrels' predators. They can be sprayed around the periphery of the garden to keep squirrels away. For a homemade repellent, try sprinkling cayenne pepper or garlic powder around plants. Planting nasturtiums, marigolds, and mustard around the border of the garden can help too, thanks to their potent aromas.

preventing and managing disease

When it comes to managing diseases in the garden, prevention is key. Fortunately, there are a number of ways to do it. In addition to planting disease-resistant crops that are adapted to your landscape, you'll want to keep tools clean, practice crop rotation, and focus on creating healthy soil. Inspect plants often so you can act quickly and proactively to keep diseases from spreading to other plants. When inspecting plants, look for stem and leaf wounds, discolored foliage, wilting leaves, leaf spots, and other signs of trouble. If you notice diseased stems and leaves, remove them and toss them in the garbage (not the compost) as soon as possible.

Keeping plants dry and mulched also helps prevent disease. Most plant diseases, especially ones caused by fungi, thrive in moist environments. Watering soil instead of foliage, pruning to ensure good air circulation, and working with plants when they're dry rather than when they're wet will go a long way in preventing disease. Mulching helps a great deal, too, as it prevents muddy water from splashing onto leaves.

Should disease arise, the challenge is first to sleuth out the cause and then find the best solution. If crops are lost to disease, the key is to understand what happened so you can do things differently next time. Gardening is just as much about learning as it is about reaping what you sow. The chart on page 152 describes some of the most common garden diseases and tips for how to prevent and control them.

DISEASE	CAUSE	SIGNS	PREVENTION/CONTROL
Blight, early	One of two different types of fungi	Affects tomatoes and sometimes potatoes. Signs include brown circular lesions up to ½ inch in diameter on leaves, marked with dark concentric circles. Usually happens before the first fruits have ripened.	Throw out diseased plants. Choose resistant varieties. Buy healthy transplants. Employ crop rotation. Water soil instead of leaves. Irrigate early in the day. Spray plants with aerated compost tea. Spray leaves with neem oil.
Blight, late	A funguslike organism called an oomycete	Affects tomatoes and potatoes. Signs include dark spots on leaves with a foul, offensive odor.	Throw out diseased plants. Choose resistant varieties. Buy healthy transplants. Employ crop rotation. Water soil instead of leaves. Spray plants with aerated compost tea. Spray leaves with neem oil.
Blossom-end rot	Lack of calcium in the plant due to low soil levels or soil that is over- or underwatered	Usually affects tomatoes; peppers and squashes are susceptible too. Signs include small, depressed, watery spots on the blossom (bottom) end of fruit. The spot starts out looking like a bruise but turns black and sunken as it enlarges.	Choose cultivars resistant to disease. Avoid over- or underwatering plants. Apply mulch. Test soil for calcium deficiency. Add organic source of calcium if needed (e.g., lime, bonemeal, finely crushed eggshells). Maintain proper pH. Avoid excessive fertilizing. Remove diseased fruit.
Clubroot	A funguslike organism that destroys plants' roots	Affects plants in the cabbage family. Signs include swollen, lumpy roots, yellowish leaves, and stunted, wilted plants.	Remove infected plants. Enrich soil with compost (instead of manure). Plant resistant varieties. Raise soil pH to at least 7.2. Disinfect tools. Rotate crops.

DISEASE	CAUSE	SIGNS	PREVENTION/CONTROL
Damping off	A fungus or mold that thrives in moist, wet conditions	Affects seedlings of many types of vegetables and flowers. Seedlings rot before or after they emerge from the soil. Plants rarely survive damping off and often an entire section of a tray of plants is killed by it.	Sow seeds in clean, sterilized trays. Provide excellent air circulation and drainage. Discard impacted seedlings and plants immediately. Avoid overwatering. Water with aerated compost tea. If germinating seedlings in covered containers, remove cover as soon as seedlings emerge from soil.
Fusarium wilt	A soil-inhabiting fungus called *Fusarium oxysporum*	Affects tomatoes, aubergines, and peppers. Leaves turn pale green to golden yellow and later wilt, wither, die, and drop off.	Plant resistant varieties. Transplant healthy-looking plants. Clean tools. Solarize soil that has been infected. Rotate crops. Remove and destroy infected plants.
Powdery mildew	A variety of closely related fungal species	Affects a wide range of plants, including tomatoes, peppers, squash, cucumbers, beans, and peas. Signs include a powdery white coating on leaves, stems, and flowers.	Remove and destroy infected leaves or plants. Choose resistant cultivars. Spray plants with neem oil or baking soda spray (1 teaspoon baking soda dissolved in 1 quart water).
Rust	A fungus that needs water on the surface of leaves to grow	Affects a wide range of plants. Signs include red, yellow, or orange spots, usually on the underside of leaves. On onions and asparagus, the tips of foliage become brownish red.	Plant resistant varieties. Keep leaves dry. Rotate crops. Remove and destroy infected leaves or parts. Dust with sulfur to treat mild cases. Spray leaves with aerated compost tea or neem oil.

extending the growing season

You might think of the growing season as the period that falls between the last and first frost dates. But with a bit of ingenuity, you can easily lengthen the season by growing crops under cover and protecting them with cloches, cold frames, greenhouses, or row covers. Choosing crops suited to cold temperatures, planting crops in succession, and mulching to keep the soil warm are other ways to extend the season. There's also the possibility of growing plants indoors with the help of grow lights or a sunny window (see page 86 for tips).

For those wishing to keep their gardens outdoors, here are a few of the most common tools for extending the growing season.

CLOCHES are short, transparent covers used to protect individual plants from autumn or winter chills. While you can buy cloches online, they can be a bit pricey. To save money you can make your own from plastic milk jugs or clear plastic bottles. Cloches need to be set deep enough in the ground to remain anchored on rainy or windy days. If temperatures rise above 50°F to 60°F, watch out for excessive heat buildup, and remove the cap to your cloche, if it has one.

COLD FRAMES are bottomless boxes placed over crops to act as small greenhouses, create a microclimate in the garden, and provide shelter and protection from wind, snow, sleet, and hail. Cold frames come in a variety of shapes and sizes, with hinged lids made of glass, fiberglass, Plexiglas, plastic, or heavy clear plastic sheeting. The lid of the box is sometimes angled slightly to increase exposure to the sun.

GREENHOUSES are a more permanent structure for protecting plants and can be heated, unheated, small, large, purchased, or homemade. Heated greenhouses allow gardeners to grow crops through the winter. Unheated ones are used primarily for starting plants in early spring and protecting them from frosts in the autumn.

ROW COVERS are perhaps the easiest way to cover plants. Fabric is either draped directly over plants or secured around a series of hoops to create a miniature hoop house. Row covers are easy to install and make for a cost-effective way to protect crops from excess cold or extreme heat. They also protect plants from pests. The fabric used is usually made from a porous polyester or polypropylene that allows sun, water, and air to pass through to plants.

HARVESTING

At long last, the crops you planted are growing leaves and flowers and fruit. If you're anything like me, you might find yourself a smidge reluctant to pluck the very plants you worked so hard to grow! Even so, come harvest time, there are few things more satisfying than making a meal with food you've grown yourself. My family and my garden really do grow hand in hand. In this section, I'll share best practices for harvesting, as well as tips for the steps that come thereafter: storing and preserving produce, saving seeds, and putting your garden to bed for the winter.

harvesting tips

One of the advantages of growing food at home is that—if you pick it at the right time—it will be tastier and more nutritious than any produce you could buy at the store. You can also harvest in a way that boosts productivity, so that you get a lot more food for a little more effort. Knowing when and how to harvest can be challenging, especially for a beginner gardener. To do it well, you need to be aware of the subtle clues indicating whether your produce is ready. Here you'll find general guidance for ensuring peak quality and maximum yields ("Plant Profiles," beginning on page 231, provides specific harvesting tips for common crops).

HARVEST OFTEN. It may seem like a paradox, but the more often you collect, the more you'll get to collect. Harvesting regularly tends to result in higher quality and better-tasting produce. It also leads to greater yields. That's because a plant's goal is to produce mature seeds that can reproduce more plants in the coming year. As soon as a plant succeeds in producing seeds, it stops producing fruit. An easy way to increase yields for plants like beans, cucumbers, peas, and summer squash is to harvest them regularly, when the fruits are young and tender. Similarly, continuously pinching back herbs encourages them to grow fuller and more plentifully.

REMEMBER, BIGGER IS NOT ALWAYS BETTER. For many vegetables, the larger the size, the poorer the quality. Vegetables like cucumbers, aubergines, green beans, okra, summer squash, and courgettes are more nutritious and flavorful when they're harvested before their seeds mature. Leafy greens like lettuce, kale, and spinach taste great when they're young and tender but become increasingly bitter with size and age.

HARVEST IN THE EARLY MORNING. Almost all vegetables do best when harvested early in the morning. Not only do they rehydrate during the night, but they also convert starches that formed during the day into sugars, making veggies harvested in the morning sweeter, crisper, and juicier. The sooner you eat them the better, as quality peaks at the time of harvest and can decline rapidly afterward.

AVOID HARVESTING IN WET WEATHER. It's best to avoid harvesting when it's raining or when plants are wet. This is especially true for disease-prone crops like tomatoes, beans, cucumbers, and squash. Handling wet plants makes it easier for disease spores to spread from plant to plant.

USE PROPER TOOLS. Many crops can be harvested easily by hand. Beans, kale, lettuce, and peas, for example, can be twisted and snapped off without causing damage to the plant. For most other crops, it's best to use a sharp knife, pruning shears, a garden fork, or other tool when harvesting, since damaged plants provide an entry point for infection and disease. For cucumbers, okra, peppers, squash, and melons, use pruning shears to cut the stem just above the fruit. For large-stemmed vegetables such as broccoli, cabbage, and courgettes, cut them off with a knife. Root veggies, like beetroot, carrots, and potatoes, should be loosened from the soil with the help of a trowel.

RESOLVE PROBLEMS QUICKLY. When you harvest, look for signs of trouble, such as yellowing leaves or rotting fruit, and resolve the problem immediately. Even if it's something you can do little about—such as blossom-end rot or cracked fruit from too much rain—it's essential to remove spent foliage and fruit so your plants can put all their energy into new, healthy growth.

storing produce

If you're lucky enough to have a bumper harvest, you may find yourself wondering how to make it last as long as possible. My grandparents had an uncanny ability to make what seemed like a modest summer harvest stretch well into the winter months. My mother can describe in great detail the way her grandmother wrapped potatoes in brown paper bags before storing them in the root cellar, and I remember watching her parents spend hours at the end of the summer preserving, pickling, and freezing fresh food. Of course, they shared a lot of their produce too—my grandfather often had neighbors line up for his tomatoes and watermelons—but if they didn't eat or share it immediately, they put it by so they could enjoy homegrown food year-round.

Here are a variety of methods for storing produce. Some are as simple as keeping vegetables in the ground until you're ready to eat them, while others require a bit more planning and preparation. Some veggies store much better than others—a good thing to know if you're hoping to eat garden produce well into autumn or winter. Potatoes, cabbage, squash, sweet potatoes, garlic, and onions are among the longest lasting crops and will keep for months if stored in the right conditions. Other veggies, such as tomatoes, peppers, and cucumbers, don't endure long in storage and are best eaten fresh.

IN THE GROUND. Perhaps the easiest way to store vegetables during the colder months is to leave them in the ground for later use. In mild-winter or mid-temperate regions, root veggies and brassicas will often do just fine, especially if you add a thick layer of mulch around the base of the plants and leave them firmly planted until you're ready to eat them. Just be sure to harvest them before the ground freezes—if you've ever tried to pluck a carrot frozen solid into the ground, you know who wins that game of tug-of-war!

IN BOXES OR SACKS. Many gardeners use special boxes specifically designed for storing produce; cardboard boxes, wooden crates, and hessian sacks work well too. One method that works particularly well, especially for carrots and beetroot, is to cut off the leaves as close to the base of the root as possible, place a layer of slightly moist sand or straw in the base of a box or crate, lay the vegetables on top so they don't touch, cover the veggies with another layer of moist sand or straw, and store them in a cool, frost-free shed or garage. Potatoes do well in hessian sacks, paper bags, or wooden crates stored in a cool, dark, dry place with ample ventilation.

ON THE SHELF. Garden catalogs sell special wooden-slatted shelf systems for storing produce, but you can easily rig up something yourself. Just place a shelf in a cool, frost-free shelter with good airflow. Crops that store well on shelves include pumpkins, winter squash, onions, garlic, cabbage, and shallots. Sometimes veggies will start to rot where they touch the surface of shelves. One way to get around that is to hang produce in netted baskets from a hook or—in the case of members from the onion family—to string them up in ropes or braids.

IN THE KITCHEN. Lots of vegetables can be processed in the kitchen for storage. Freezing is a relatively easy way to store fruits and vegetables that deteriorate quickly, such as berries, peas, beans, asparagus, and corn. Some types of vegetables will need to be blanched first, which involves boiling whole or cut-up pieces of vegetables for 1 to 2 minutes and then immediately placing them in ice-cold water to stop the cooking process. Blanching slows the ripening process of veggies and improves their flavor, color, texture, and nutrient retention. Frozen vegetables will keep for up to one year when packaged well. Store them in plastic or silicone bags that are durable, leakproof, and resistant to moisture and vapor. Tempered glass jars specifically designed for freezing are also suitable. In addition to freezing, you can pickle, ferment, can, or dry your produce. If you don't own a dehydrator, you can dry a lot of foods in an oven set to the lowest temperature setting.

preserving herbs

If you've ever grown herbs, you're probably accustomed to having more than you could possibly use in one season. Luckily, you can dry them and enjoy the perks of summer all year long. If you harvest, dry, and store them properly, herbs can retain their flavor and medicinal value for up to a year. Here are a few tips for preserving them.

HARVEST

It's important to harvest herbs at the right time. Pick them after their flower buds appear but before they open, when the plant has the highest concentration of essential oils. It's also best to pick them in the morning, after the morning dew has evaporated but before the sun gets too hot. Harvesting frequently helps promote new growth.

CLEAN

Avoid washing herbs until just before you plan to use them, as dampness promotes the growth of yeast and mold. Instead, remove old, dead, or wilted leaves by hand and clean the leaves and stems by shaking or brushing away debris. If you use organic gardening practices you won't have to worry about washing off chemicals.

FREEZE

Freezing is an easy way to preserve herbs with large leaves and a high moisture content, such as basil, mint, and coriander. Simply strip the leaves from the stems, arrange them in a single layer on a baking sheet, and place them in the freezer. Once they are frozen, transfer them to an airtight container and store them in the freezer until you're ready to use them. This method keeps leaves separate so you can take them out one at a time. Herbs with small leaves, such as rosemary and thyme, should be frozen on their stems.

Another way to preserve high-moisture herbs involves freezing them into ice cubes. Just remove them from their stems and pack ice cube trays with chopped or whole leaf herbs, cover them with water or olive oil, and pop them in the freezer. Once they're frozen, transfer the cubes to an airtight container and store them in the freezer.

LOW-MOISTURE HERBS	HIGH-MOISTURE HERBS
bay	basil
dill	borage
lemon balm	chives
lemon verbena	coriander
marjoram	lemongrass
oregano	mint
rosemary	parsley
sage	rocket
tarragon	
thyme	

DRY

Air-drying works best for herbs with a low moisture content. Low-moisture herbs should be dried in the dark with good air circulation and temperatures below 110°F. You can use a food dehydrator (set between 90°F and 110°F), flat baskets, clean window screens, or clothes-drying racks. (Although they might seem like a good shortcut, microwaves or ovens should be avoided, as they tend to cook the herbs and destroy their flavor, oils, and color.) You can also tie up herbs in bundles of four to six stems and hang them upside down to dry. Herbs have finished drying when you can easily crumble them in the palm of your hand. To prevent contamination and oxidation, take down the bundles and store them as soon as they're dry.

STORE

When it comes to storing dried herbs, whole leaves and seeds retain their oils and flavor better than crumbled ones. Either way, store them in airtight glass jars out of direct sunlight and away from heat. Label jars with the date and contents. You can tell whether an herb is still valuable by its look, smell, and effectiveness. As herbalist Rosemary Gladstar says, it should look, smell, and work just as it did on the day it finished drying. When using dried herbs in a recipe, keep in mind that dried herbs are more concentrated than fresh ones. Generally speaking, if a recipe calls for 1 tablespoon of fresh herbs, you can substitute 1 teaspoon of crushed dried herbs.

saving seeds

For some gardeners, growing food and flowers is about connecting with the full life cycle of plants. While many gardeners grow crops for food, some also grow them for seed. From saving money to preserving food culture, there are lots of good reasons to grow, collect, and save your own seeds.

SAVING SEEDS SAVES MONEY. Buying seeds isn't all that expensive, compared to buying starts, but the cost of planting your garden can add up fast, especially if you have a large plot to sow.

SAVING SEEDS PRESERVES GENETIC DIVERSITY (AND YOUR FAVORITE VARIETIES). The commercial seed industry is dominated by just a few big companies that focus on a small handful of crop varieties. Saving seeds has become essential to preserving genetic diversity. It's a great way to ensure you'll have access to the varieties you've come to know and love. You never know when a seed catalog might discontinue a favorite heart-shaped tomato or perfect pickling cucumber in order to make and market new varieties. Your only guarantee for having access to your favorite varieties is to save their seeds yourself.

SAVING SEEDS HELPS GARDENERS CONNECT WITH THEIR ANCESTORS. Until just a few decades ago, it was common for gardeners to work alongside their families and communities to save seeds for future generations. Saving seeds from homegrown vegetables is starting to make a comeback, as gardeners are becoming increasingly interested in making sure future generations can enjoy the same varieties of tomatoes and great-tasting beans their great-grandparents grew. To save seeds is to preserve history and culture; if not for the gardeners who devote themselves to growing and saving seeds, heirloom varieties would probably go extinct.

SAVING SEEDS HELPS POLLINATORS. Pollinators are vital to heathy, stable food supplies and the varied and nutritious diets we need and enjoy. Pollinators are also in decline. Allowing crops to go to flower and go to seed provides an invaluable food source for bees, butterflies, and beetles.

SAVING SEEDS PROMOTES SELF-RELIANCE. In times of trouble, or when seeds are hard to find, it's reassuring to know you can fall back on your own personal stash. Saving seeds allows you to have greater control of your food supply.

SAVING SEEDS ADAPTS PLANTS TO YOUR GARDEN AND IMPROVES THE QUALITY OF CROPS. To save seeds is to participate in the process of evolution. When you save seeds from your healthiest, tastiest, most robust plants, and continue to do so season after season, you gradually make improvements to the quality of those seeds.

SEEDS

tips for saving seeds

Saving seeds isn't hard. Once you know the basics, it's easy to save them from crops like tomatoes, lettuce, and beans. Saving seeds from crops like carrots and squash is a bit more challenging, but with determination it can be learned and mastered. Before diving in, it helps to understand different types of seeds, as well as best practices for harvesting, drying, and storing them. Here are a few tips to help you get started.

GROW OPEN-POLLINATED SEEDS.

Make sure you save seeds from open-pollinated varieties and not from so-called F1 hybrids. Open-pollinated seeds are those that, if properly isolated from other varieties in the same plant species, will produce seeds that are genetically "true to type," meaning they'll look similar to their parents. Heirloom seeds are a type of open-pollinated seed that have a history of being passed down within a family or community. An heirloom variety must be open-pollinated, but not all open-pollinated seeds are heirlooms.

If open-pollinated varieties are allowed to cross with other varieties within the same species, the resulting seed will be a hybrid. This can happen naturally or artificially. An F1 hybrid is the first-generation offspring from a deliberate cross of two different parent plants. These plants are bred to produce desirable traits, such as disease resistance, outstanding vigor, and high productivity. Although the first generation of plants produced from F1 seeds will produce these traits, the second generation will not be true to type. What that means is that if you plant the seeds of F1 hybrid plants, you can't be sure what qualities they will have.

START WITH SELF-POLLINATING PLANTS.

Seeds are the product of pollination. Some plants self-pollinate, meaning the plant can fertilize itself. Other plants cross-pollinate and need pollen from another plant in order to set seed. The easiest way to save pure seeds—that is, to ensure seeds that grow true to type—is to collect them from plants that self-pollinate.

SELF-POLLINATING CROPS	CROSS-POLLINATING CROPS
beans	broccoli
lettuce	corn
peas	cucumbers
peppers	leeks
tomatoes	melons
	onions
	radishes
	squash

START WITH ANNUALS INSTEAD OF BIENNIALS.

It's easier to collect seeds from annual crops than biennial ones because annual crops go to seed the same season they're sown. Biennial crops like carrots and beetroot, on the other hand, have to be kept alive and in good condition through the winter in order to produce flowers and seeds during their second year of growth.

GROW A SINGLE VARIETY OF CROSS-POLLINATING CROPS.

Seed saving gets a little complicated with cross-pollinating crops, such as corn, cucumbers, and squashes. To save pure seed, you need to prevent cross-pollination between two different varieties of the same species. Seed savers use various strategies to isolate varieties, including distancing, caging, and bagging, but the easiest way to prevent cross-pollination is to plant only one variety of these species at a time. Then you can watch insects pollinate flowers without worrying they'll sully the purity of each crop's seeds. Can you imagine herding a bee to the right flower? Sounds exhausting!

SAVE SEEDS FROM THE BEST PLANTS.

Saving seeds allows you to choose which plants get to reproduce. Observe your plants well, taste their produce, and save the seeds from the best performers. If you want tomatoes that ripen early, save seeds from the first fruits that ripen each year. If you want disease-resistant squash, save seeds from the fruits that matured unscathed. Eventually, your seeds will have the traits you most desire, while being well adapted to growing in your particular climate.

TAKE EXTRA-GOOD CARE OF CROPS YOU'RE GROWING FOR SEED.

If you know you're growing a plant specifically for its seed, take care of it as well as—if not better than—plants you're growing to eat. After all, it will become the blueprint for future crops. Start with good-quality seeds and provide them with rich soil, adequate water, and protection from pests and disease.

HARVEST SEEDS AT THE RIGHT TIME.

Seeds are ready to save when they are fully mature, which does not always coincide with when a plant or fruit is ready to eat. It's important to get the timing right, as seeds picked too soon won't germinate. With the exception of tomatoes, seeds should be harvested as late in the season as possible.

CLEAN SEEDS PROPERLY.

Seeds must be cleaned properly to prepare them for storage. The methods for cleaning dry seeds are different than those for cleaning wet seeds. Dry seeds, such as those for lettuce and radishes, are usually mixed with chaff—pieces of the flower head, stem, and leaves. You can separate the seed from the chaff by hand or, if you have a lot of seeds, you can speed up your work by using mesh of the appropriate size. Winnowing is another option and involves placing seeds in a large flat basket and repeatedly tossing them into the air. If there's a light breeze, the chaff will blow away while the heavier seeds fall back into the basket. If there's no natural breeze, you can create an artificial one by using a fan.

Cleaning wet seeds, such as those for squash and tomatoes, is a bit messier and—in the case of tomatoes—more involved. To clean wet seeds from the squash family, simply place the seeds in a colander and rinse them with water until the wet, sticky film washes away. (See the tutorial opposite for how to clean tomato seeds.)

DRY SEEDS PROPERLY.

Cleaned seeds should be dried in an airy place at room temperature. To speed up the drying process, place a small fan on low speed nearby so that it blows gently on the seeds. To test whether a seed is completely dry, place one end between your teeth and bend the other end with your fingers. A dried seed will break in half. For larger seeds like beans, place one on a hard surface and hit it with a hammer. If it's dry, it'll shatter into pieces.

STORE SEEDS PROPERLY.

To preserve their viability, keep seeds away from heat, humidity, sunlight, and hungry rodents. Put them in sealed paper envelopes, labeled with their crop and variety name, as well the month, day, and year they were collected. Place the envelopes in an airtight container that will keep out moisture (and rodents!), such as a glass jar or plastic storage container. For added protection against humidity, place milk powder or silica gel at the bottom of the container. Keep the container of seeds in a cool, dry, dark place, such as a basement or closet. A refrigerator is a safe place to keep seeds, too—just be sure to store them in airtight containers. Allow the refrigerated container to warm to room temperature before opening it to prevent moisture from condensing on the seed packets.

how to clean and save tomato seeds

Cleaning tomato seeds requires an extra step that uses fermentation to remove the gelatinous material surrounding the seeds. It's not the most straightforward process—and it's stinky, too—but if you want to save the seeds of a special variety of tomato, it's worth the effort.

MATERIALS

1 fully ripe, healthy tomato
wide-mouth mason jar
cheesecloth
rubber band
fine-mesh sieve
plate, screen, or towel for drying

DIRECTIONS

1. Wash the tomato and cut it in half. Scoop the pulp, seeds, and juice into the jar.

2. Cover the jar with cheesecloth and secure it with a rubber band.

3. Place the jar in a warm place (about 70°F) and out of direct sunlight for several days. Stir once or twice a day.

4. After 2 to 5 days, the viable seeds will sink to the bottom of the jar. Mold, liquid, and unusable seeds will rise to the top. Continue to stir the fermenting juices once or twice daily. This prevents the buildup of mold, which is not harmful to the seeds but may discolor them.

5. After 5 days, pour off the mold, liquid, and seeds that floated to the top, but not the viable seeds that sank to the bottom. Transfer the viable seeds to a fine-mesh sieve and rinse them until they're clean.

6. Spread out the wet seeds in a single layer on a plate, screen, or towel. Allow them to dry for 7 to 14 days in an airy spot at room temperature.

7. Store the seeds in a labeled envelope in a sealed container until you're ready to plant them.

preparing the garden for winter

My love for gardening knows no bounds. But come autumn, as the leaves turn, the days shorten, and cold winds start to blow, I'm as ready as my garden for a little break. There's something about living in step with the seasons that feels right—that connects me to the earth and, if I'm being honest, to the best version of myself.

So, while in spring I'm all aflutter with my seedlings and in summer I'm buzzing nearly nonstop among my beds, in autumn I'm taking great care in covering over and tucking in my garden for a long winter's rest. It is deeply satisfying, knowing I can retreat indoors and spend those cold months sipping hot tea and planning for next year, already anticipating the great delight of watching a new garden spring to life.

Not to get too metaphysical, but I like to think a good gardener lives as cyclically as her garden, replete with little rituals for the turning of each season. The rituals that prepare my garden for winter include clearing out spent and dying plant debris, leaving habitat for overwintering beneficial organisms, testing and amending the soil, protecting against erosion, helping perennials survive winter, and cleaning and storing tools and containers.

TIDY PLANT DEBRIS.

Once you've harvested most or all of your veggies and seeds, remove perennial weeds and diseased, pest-infested plants from the garden. Don't toss them in the compost—otherwise you'll introduce pests and diseases to the garden come spring. Leave hardy or half-hardy plants in the garden and continue to harvest them as long as possible.

LEAVE SOME HEALTHY PLANTS.

Although you might be inclined to cut, pull, and trim every plant in the garden, there are good reasons to keep roots and a few healthy shoots in place. Not only will root networks decompose slowly and feed soil life throughout the winter, but hollow stems and leaf litter will provide winter shelter and habitat for beneficial insects and other creatures.

TEST AND AMEND THE SOIL.

Autumn is a great time to test your soil. Once you get your results back and know the pH levels, nutrient levels, organic matter content, and general health of your soil, you can amend it as needed to get it in tip-top shape for spring.

COVER THE SOIL.

Whatever you do, try not to leave your garden bare over the winter, as rain, wind, and snow can cause erosion. Instead, top it off with organic compost to replenish nutrients and cover it with autumn and winter transplants (in mild-winter areas), a layer of mulch, or a cover crop.

PROTECT HERBS FOR THE WINTER.

Some perennial herbs are winter hardy, including chives, lavender, mint, oregano, and thyme. In most climates (USDA hardiness Zone 5 or below), these plants just need a good pruning—down to a height of 4 to 6 inches—and a 3- to 6-inch layer of mulch. Rosemary, bay laurel, and lemon verbena need a little extra support during the winter months. They should be cut nearly to the ground after the first hard frost, then covered with 4- to 6-inches of mulch.

Tender herbs like basil or parsley can be potted and brought indoors until spring.

CLEAN AND STORE TOOLS AND CONTAINERS.

Tools, supports, and containers will degrade if exposed to the elements and will do better stored in a protected location like a basement, garage, or shed. Empty the containers and store potting soil from healthy plants in clean bags or lidded bins so it can be reused in the spring. To prevent the spread of pathogens, clean pots with soapy water and sterilize them with hydrogen peroxide (diluted with water at a ratio of 1:10). Remove hoses from taps and drain them before hanging them indoors.

PLAYING

I delight in gardening. I can't think of a time when a day spent in my garden felt like a day spent working. As a child, I played in my grandfather's garden. As an adult, I now play in my own. Ask anyone who knows me and they'll tell you there are few things I enjoy more than diving headfirst into an artsy project somewhere between the pea shoots and the bean sprouts.

Sometimes my children join in the fun. From their youngest days I looked for ways to introduce them to growing food and sharing in the joy I feel when tending my crops. Watching them putter around the plants, I can't help but think there might be no better nursery than a garden.

To my daughter, the little nook in the fence beside a raised bed full of cabbages is a refuge for traveling fairies, and she tends to it as assiduously as her patch of radishes and cucumbers—creating fairy beds with cotton, wool, and twigs and leaving snacks, ornaments, and keepsakes. On warm summer evenings, she often bolts outside after dinner to share a few crumbs of her dessert and drop off messages she's written on tiny scraps of paper.

My son is a teenager now, and so these days his version of fun is to try to water the rest of us every bit as much as he waters the plants. But it's enough for me to know he still likes to venture back there, wandering among the beds, recognizing different plants, knowing a little bit about how best they grow, and just feeling comfortable, even happy, in their presence.

In this section are projects and ideas for adults and children alike to play in the garden. All are activities I do myself and, when they're interested, with my children—including building a living teepee out of bamboo and pole beans, folding origami seed envelopes, growing a hessian sack of potatoes, making garden markers out of clay, rolling newspapers into pots, and enjoying herbs and edible flowers in biscuits, cookies, ice lollies, and lollipops.

Never forget that even the basics of gardening can give great joy—sowing seeds, watering seedlings, transplanting plants, and harvesting crops. The very act of gardening creates a connection to the natural world. The more we feel the world is nurturing us as we nurture the world, the more, I think, we'll all be as healthy as our planet ought to be. To play in your garden is to draw on a deep well of inspiration, lighting the way while seeking to safeguard the environment and pursue sustainable solutions.

Maybe most of all, playing in the garden connects me to my own childhood, to my parents and grandparents, every bit as much as it allows my children to connect to me. As Robin Wall Kimmerer said: "This is really why I made my daughters learn to garden—so they would always have a mother to love them, long after I am gone."

HUNTER
HUNTER

3/22

newspaper seedling pots

Making newspaper seedling pots is a wonderful way to include children in the garden. They're easy for little hands to roll up with the help of a seed pot form or a small tin can. They're also free and sustainable, requiring nothing more than newspapers from the recycling bin. Unlike plastic seed trays and pots, they're completely biodegradable and can be planted directly in the ground to protect seedlings from transplant shock. Over time the newspapers will decompose, just as they would in a compost bin. When selecting your materials, be sure to use standard black-and-white newspaper with nontoxic soy-based ink.

MATERIALS

sheet of newspaper (approximately 22 by 12 inches)
seed pot form or 6-ounce can
seed-starting mix
seeds of your choice
waterproof tray

DIRECTIONS

1. Fold the newspaper in half crosswise so you have a two-sheet-thick stack. Cut into thirds lengthwise to create three strips, each having two layers.

2. Place the can on its side on the newspaper, leaving 1 to 2 inches of newspaper hanging past the bottom of the can. Tightly roll the newspaper around the can until you reach the end of the newspaper.

3. While holding the end of the newspaper with one hand, use your other hand to fold the 1 to 2 inches of newspaper hanging past the bottom of the can down over the bottom of the can, working your way around it until all the edges are crimped down. Flip the can right-side up and press it against a table to tighten the folds on the bottom of the pot.

4. Slide the can out of the newspaper pot. Don't worry if the pot feels flimsy; it will become surprisingly sturdy once it's filled with soil and water. Repeat with the other two newspaper strips.

5. Fill the newspaper pots with the seed-starting mix, plant your seeds, and store the pots in a waterproof tray. Water seeds daily.

6. Once your seedlings are ready to be transplanted, plant the entire newspaper pot in the soil.

TIP: *Biodegradable pots can also be made from toilet paper rolls or the empty rinds of grapefruits, oranges, and lemons. Both can be transplanted directly into the soil.*

seed tape

If you've ever snooped around a garden center, you may be familiar with pre-made seed tape. If not, it's essentially a long strip of biodegradable paper with seeds embedded in it. The seeds are spaced the correct distance for growing, to reduce seed waste and eliminate the chore of thinning seedlings. Seed tape is especially helpful to young gardeners who sometimes struggle to separate and sow teeny-tiny seeds like carrots and onions. If you have a stash of seeds at home, you can make seed tape yourself, using common household supplies.

MATERIALS

scissors
sheet of newspaper
all-purpose flour
lukewarm water
seeds
pencil
ruler
small, clean paintbrush
rubber bands, paper clips, or twine
mason jar with lid

DIRECTIONS

1. Using scissors, cut the newspaper into 2-inch-wide strips.

2. Create seed glue by mixing 1 tablespoon flour and 2 tablespoons lukewarm water. Add more flour or water as needed to create a mix similar in consistency to PVA glue or papier-mâché paste.

3. Label the bottom edge of the strip with the type of seed and planting depth instructions. Look at the back of the seed packet for instructions on seed spacing. Using a ruler and pencil, mark the newspaper strips at the recommended intervals.

4. Using the paintbrush, apply a dab of seed glue on each mark. Place two seeds on each spot of glue.

5. Allow the strips of seed tape to dry completely, then roll them up and secure the roll with a rubber band, a paper clip, or twine. If you do not plan to use the seed tapes immediately, store them in an airtight glass jar in a cool, dark place until you're ready to plant them.

6. To plant your seed tape, dig a trench in your garden bed, to the depth indicated on your seed packet. Then unroll the tape, seed side up, and set it in the trench. Cover the tape with compost or soil and water it evenly.

Equity Funds

origami seed envelopes

Although you can buy envelopes for storing seeds, it's fun and easy to make them yourself. It's an especially great project for children who love to craft as much as they enjoy gardening. Although origami paper is a natural choice for this activity, you can also upcycle newspapers or make thrifty use of old seed catalogs or magazines. Once you've stuffed your envelopes with seeds, be sure to store them in a cool, dark place until you're ready to sow them.

MATERIALS

scissors	**seeds**
scrap paper	**label stickers**

DIRECTIONS

1. With scissors, cut a 6-by-6-inch square piece of paper. Fold it diagonally into a triangle and position it so that the longest side is facing you.

2. Fold the bottom right corner to meet the center of the opposite side of the triangle. The top edge of the fold should be parallel with the longest side facing you.

3. Fold the bottom left corner to meet the center of the right side of the triangle.

4. There should be two triangular flaps at the top of the envelope. Tuck the top triangle into the pocket created by steps 2 and 3.

5. Fill the envelope with seeds, then tuck the remaining triangle into the pocket. Label the envelope with the name of the seeds and the date they were collected.

1.

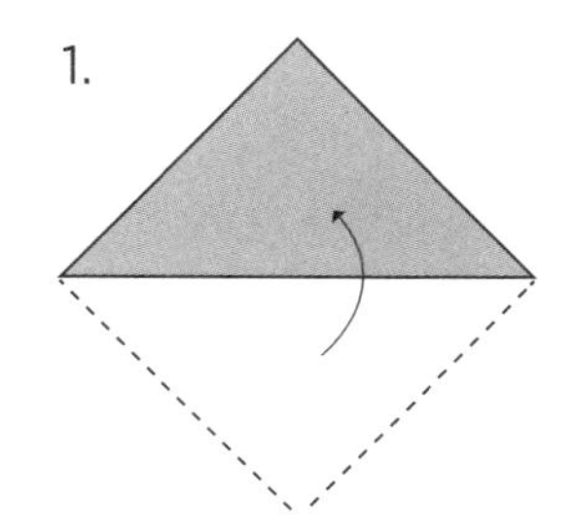

2.

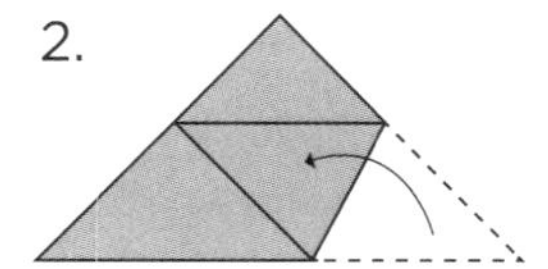

3.

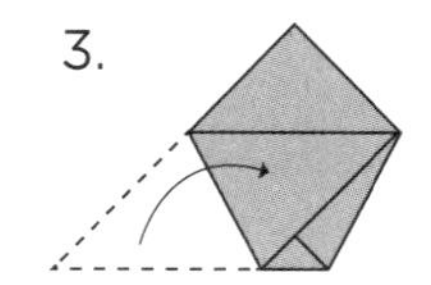

4.

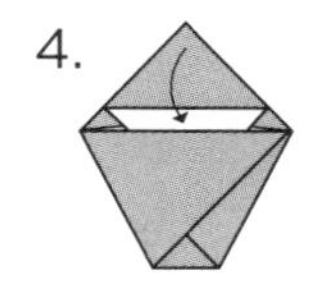

5.

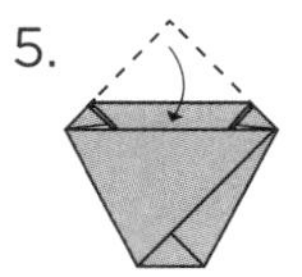

stepping-stones

Homemade stepping-stones are a fun way to personalize the garden, especially for kids. They're easy to make, too—just mix concrete, pour it into a mold, let your children decorate it, and allow it to dry. If you're making multiple stones, your best bet is to buy a bag of concrete at the hardware store. For smaller projects, you can purchase stepping-stone concrete from your local craft supply store.

To decorate the stones, look around your house and garden for ideas. Marbles, gemstones, sea glass, river stones, pea gravel, and seashells all make beautiful embellishments. To make impressions, use flowers, pine cones, acorns, or other natural objects. Some children like to make a handprint or footprint impression. As for molds, you can find them at craft stores in myriad shapes and sizes, or you can use anything that will hold the concrete's shape while it dries. Old cake tins or baking trays are my go-to molds for simple round or square-shaped stones. Since the concrete can be hard to scrape off the mold, use something you don't mind mucking up or repurpose old tins from a thrift store.

MATERIALS

newspapers or drop cloth
concrete molds
cooking spray or oil
bucket
quick-setting concrete
decorative objects

DIRECTIONS

1. Protect your work area by covering it with newspapers or a drop cloth. If the weather is nice, consider making your stones outdoors.

2. Grease your molds with cooking spray or oil.

3. Mix the concrete in the bucket according to the instructions on the bag. Pour the concrete into the molds and wait for it to settle until the top is flat.

4. After 5 minutes, begin adding decorative objects to the top of the concrete and/or make impressions with your hands, feet, or interesting objects.

5. Allow the concrete to dry for 24 to 48 hours, then remove it from the molds and set your stones in the garden.

THYME

clay garden markers

These garden markers are every bit as cute as they are practical, and they make a great sensory activity for gardeners of all ages (including adults!). If your children are under the age of seven, they'll probably need some help, but they'll love rolling the clay into balls, flattening it out with a rolling pin, and labeling the markers with rubber stamps. If you wish to avoid plastic, stay away from polymer-based clay, as it's made from polyvinyl chloride (PVC), and is difficult to dispose of and recycle. Instead, look for an earth-based air-dry clay. I use a brand by Das that is nontoxic and comes in white or natural terra cotta.

MATERIALS

air-dry clay | **rimmed baking sheet**
rolling pin | **rubber alphabet stamps**
butter knife | **oven**

DIRECTIONS

1. Break off pieces of clay from the original form and roll them into 1¼-inch balls.

2. Roll each ball into a long, even rope about 5 to 5½ inches long. Using the rolling pin, flatten the rope into a flat rectangle approximately ¼ inch thick.

3. Use the dull side of a butter knife to trim one end straight across and the other end into a point (like a stake). Trim the long sides to make them straight. Shape the markers with your hands, if needed, and transfer them to a baking sheet.

4. Using rubber stamps, indent the names of the veggies and herbs into the garden markers. Place the baking sheet in the oven and bake the clay according to the directions on the package.

5. Allow the markers to cool.

6. When you're ready to use the markers, soften the soil with a trowel and gently press the markers into the loosened soil. At the end of the gardening season, clean the markers and store them in a dry place.

upcycled chair planter

When it comes to finding containers for plants, I like to think outside the box. A rustic colander turned hanging basket, an old farmhouse sink filled with soil and flowers, a forgotten chest of drawers converted into a vertical garden—all add character to a garden. Upcycling vintage chairs into flower planters is one of my favorite garden crafts. In addition to adding greenery, they add interest and height to the garden. Children love to help too! Bring them along to the thrift store and let them pick out a chair, then give them paint and a brush and invite them to decorate away. They can also help choose flowers and herbs for the planter.

When shopping or searching for the right chair, your first priority should be to find one with a removable seat. While you could technically cut a hole in a solid wooden seat, it would make this project unnecessarily difficult. It's much easier to look for a chair with a wicker, fiber, or removable seat that can be cut or popped right out of the base. If, like me, you gravitate toward vintage, be sure to choose a chair that is not covered in lead-based paint. Although the chippy, weathered look can be appealing, the last thing you want is to introduce toxic substances into your garden.

MATERIALS

old chair with removable seat
paint or other decorations
work gloves
wire cutters
chicken wire
floral wire
coco fiber planter liner
potting soil
flower or herb plants

DIRECTIONS

1. Remove the chair's seat. Paint or decorate the chair as you wish.

2. Wearing work gloves and using wire cutters, cut a section of chicken wire to fit inside the open seat of the chair. The wire should be large enough to form an 8- to 12-inch-deep basket that can extend over the chair's edges.

continues

upcycled chair planter, continued

3. Place the center of the chicken wire inside the open seat of the chair. Press the wire down into the seat to create a basket 8 to 12 inches deep. Bend the excess wire over the edge of the chair. If there is more wire than you need, cut away some of it, but make sure to leave enough to secure the basket to the chair.

4. Using floral wire, secure the basket to the seat frame of the chair. Twist and tie the floral wire to the frame every 3 to 4 inches. Trim away any excess chicken wire or tuck it neatly under the frame of the seat.

5. Place the coco fiber planter liner inside the chicken wire basket. Pull and spread the edges of the coco fiber liner over the seat frame of the chair to conceal the chicken wire.

6. Place a layer of potting soil inside the coco fiber liner. Transplant your plants and cover them with another layer of potting soil. Set your chair planter out in the garden and water it well.

TIP: *If you want to conceal the chicken wire completely, plant herbs, flowers, and veggies that cascade over the rim of the basket.*

sack of potatoes

Growing potatoes is a perfect beginner gardening project, especially for children. Potatoes are nutritious and tasty, easy to grow, and a joy to harvest. If you have plenty of room in the garden, you can grow potatoes in rows for a sizable crop. Otherwise, if space is limited, growing potatoes in a sack is an excellent option. Keep in mind that potatoes don't do well in hot weather, so be sure to plant them in early spring, when daytime temperatures are in the 60s. A good rule of thumb is to plant them three to four weeks before your last frost date and to place your sack in a spot that gets more than six hours of sunlight a day.

MATERIALS

3 seed potatoes
paring knife
1 hessian sack, 2 feet wide x 4 feet high
potting soil
compost
trowel
watering can
cardboard box
paper bag

DIRECTIONS

1. Cut the seed potatoes into 2-inch pieces, making sure each piece has one or two eyes. Set the pieces cut side up on a counter at room temperature, and allow the pieces to dry out for a few days to keep them from rotting once planted. If the potatoes' eyes start to grow shoots during this time, don't worry—this is totally fine.

2. When you're ready to plant the potatoes, place the hessian sack outside where you want your potatoes to grow. Roll down the edges of the sack to create a thick, supportive container about 10 inches tall.

3. Fill the hessian sack three-quarters of the way with potting soil and top off the rest with compost. Using your trowel, mix the soil and compost well.

4. Place the potato pieces on the soil, about 4 inches apart with the eyes facing up. Push the potatoes 2 inches into the soil and cover them completely with a layer of compost.

continues

sack of potatoes, continued

5. Water the top of the soil lightly so that it's moist but not sopping wet.

6. Check your potatoes daily and water them if needed, continuing to keep the soil moist but not too wet. After two to three weeks, the seedlings should start to emerge.

7. When the potato plants are 2 to 3 inches tall, add more potting soil and compost to the sack to create a mound around the stem of the potatoes, but leaving the leaves unburied. This is called earthing up.

8. As your plants grow, continue earthing up the potatoes so that their stems are covered. Unroll your bag as needed to make room for more potting soil and compost.

9. When you start to see blossoms on your plants, that means potatoes have started to grow below the soil's surface. Allow the plants to continue to grow until the leaves and stems wilt and turn brown (for most potatoes, this will take about 90 days from when you plant them).

10. Harvest the potatoes by carefully breaking up the soil around the plants with a trowel. Then use your hands to dig up the potatoes.

11. Place the potatoes in a single layer in a cardboard box and store them in a cool, dry, dark place to allow them to cure, or dry, for up to two weeks.

12. Once the potatoes have dried out, place them in a paper bag, roll up the bag tightly, and store it in a cool, dark, and dry place.

13. When you're ready to eat your potatoes, wash and scrub them first, then enjoy!

ROOTS IN BOOTS

Turn worn-out wellington boots into containers! Simply drill drainage holes in the soles, cover the holes with screening, fill the boots to the rim with potting soil, and tuck seedlings into the soil. We like to plant carrots, parsley, and radishes in our boots, but flowers work nicely too. Water well and place your boots on a flat piece of land in full sun. (See the photo on page 177.)

basket of strawberries

Strawberry plants are a great first crop—they spread easily, grow well in containers, and are fun to harvest (and eat!). Some gardeners put them in pots, but another fun way to grow them is in baskets. Just be sure to use something you don't mind getting dirty! To save money and resources, consider upcycling a secondhand basket from a thrift store. If you have kids, this is a great way to include them in the garden. They'll love having ownership of their very own transportable container of berries.

MATERIALS

scissors
1 13-gallon trash bag
1 basket at least 6 inches deep and 12 inches wide
trowel
potting soil
compost
2 or more strawberry plants
organic fertilizer

DIRECTIONS

1. Use scissors to cut 1-inch slits randomly along the bottom 3 inches of the trash bag.

2. Place the trash bag inside the basket and press the sides of the bag against the inside walls of the basket. Cut the top of the bag about an inch below the rim of the basket.

3. Use the trowel to fill the basket with potting soil up to 3 inches from the top of the basket. Add a 1-inch layer of compost and mix it into the potting soil.

4. Transplant the strawberry plants 6 inches apart. If your basket is more than 12 inches wide you can fit more than two plants. Add more potting soil, if needed, to conceal the trash bag.

5. Set or hang the basket in a sunny spot. Water the strawberries daily, avoiding wetting the leaves as much as possible. Feed the strawberries with an organic fertilizer every 10 to 14 days.

6. Harvest strawberries when they are soft and red.

homemade scarecrow

Scarecrows aren't just for fun. They're practical, too! They've been used for centuries to discourage birds from disturbing and feeding on seeds and crops. Building a scarecrow is a wonderful family activity, as well. Parents can build the support frame while kids can pick out clothes for their new stuffed friend. When my children were little, they loved accessorizing our scarecrow—every time I turned around, it was wearing a new hat, bandana, or handkerchief!

MATERIALS

3-foot long 2x4 wooden stake
5-foot long 2x4 wooden stake
twine
scissors
1 yard of hessian
straw or hay
old trousers and shirt
boots (optional)
hammer
old hat (optional)

DIRECTIONS

1. Build the frame. Create a cross with your 2x4s by placing the shorter 3-foot-long stake horizontally across the longer 5-foot-long stake. The shorter piece will be the scarecrow's arms so place it where you want them to be. Tie the wood together with twine.

2. Make the head. With scissors, cut the hessian material into an 18-by-18-inch square. Place several handfuls of straw or hay in the middle of it, then gather the ends and bring them together. Bunch up the hessian beneath the hay and shape the bundle into a round head. When you are happy with the shape, tie the hessian below the head with twine.

3. Dress the scarecrow. Cut a hole at the base of the crotch and slide the trousers over the vertical stake. Guide the shirt through, over, and around the vertical and horizontal stakes. Tuck the bottom of the shirt into the trousers. Tie the bottom of the trouser legs with twine. Stuff each trouser leg and the arms and body of the shirt with straw or hay. Tie the ends of the arms of the shirt with twine.

4. Find a spot for your scarecrow in the garden. Hammer the vertical stake into the ground until your scarecrow can stand on its own. Slip the boots over the trouser legs.

5. Add the head. Slip the hessian head over the top of the vertical stake. If needed, trim any excess hessian below the tie. If you have an old hat, throw it on top of the head.

bean teepee

Pole beans need vertical supports to grow upward. While simple stakes and trellises will work, a bean teepee adds an element of surprise and wonder to the garden. For children, it's an irresistible hiding spot during the heat of summer. We built bean teepees when my children were younger and found them every bit as fun as the blanket forts that defined my childhood. You won't need much to create them—just a set of bamboo poles, some rich compost, and a packet of pole bean seeds. If you want, you can add other climbing plants to your teepee, such as scarlet runner beans, miniature pumpkins, birdhouse gourds, climbing nasturtiums, or vining black-eyed Susans.

MATERIALS

6 to 8 bamboo poles, 6-10 feet long
long rope or hose
garden hoe
shovel
compost
hemp or jute twine
pole bean seeds
other climbing seeds (optional)

DIRECTIONS

1. Choose a location for your teepee that gets 6 to 8 hours of sunlight a day. Loosely assemble the bamboo poles in a teepee shape, but don't tie them together just yet. If you plan to use the teepee to create a getaway for children, be sure to create a circle at least 3 feet wide. Mark the border with a rope or hose, then disassemble the poles and set them on the ground.

2. Determine where you'll put the doorway for the teepee and mark off that area with a couple of rocks or bricks. Using the hoe, remove any sod or vegetation from the teepee's outer circumference, then use a shovel to dig a 3-inch-deep trench and fill it with compost.

3. Position your teepee poles, leaving a gap wide enough for a doorway. Press the poles firmly into the ground and tie the tops together with twine.

continues

bean teepee, continued

4. To help the beans fill out the teepee, run twine horizontally across the poles every 12 inches or so, from top to bottom. This way, the plants can grip the poles and the twine and create a full, dense teepee.

5. After your last frost date, plant two to four bean seeds at the base of each pole. You can also add seeds for other climbing plants, such as morning glories, between the poles. Follow the directions on the seed packets to determine the best spacing for the particular plants you choose.

6. Water the seeds daily. As the plants grow, guide them up the poles and supportive twine. If the beans aren't climbing on their own, attach the tendrils to the twine. Although they may seem slow at first, you'll be surprised how quickly they take off once they get started.

TIP: *You can use this same technique to create smaller teepees purely for support. Adjust the diameter of the circle as needed to fit your space.*

gourd birdhouse

Every year, my children protest that we should grow a garden for all the wild animals, and every year I convince them that that's precisely what we do! We grow wildflowers for pollinators, milkweed for monarchs, sunflowers for squirrels, and gourds for birds. In some cultures, hard-shelled gourds are grown for crafting bowls and spoons, but at our house, we grow them to make birdhouses . . . from scratch!

MATERIALS

bottle gourd seeds
pruning shears
newspapers
scrub brush
hole saw
drill
wire, twine, or leather cord
paint or wood-burning tool (optional)

DIRECTIONS

1. Grow the gourds according to the directions on your seed packet. Be sure to choose a location with full sun and fertile soil, near a trellis, fence, or arbor that can support them as they grow. Gourds have a long growing season and transplant poorly. If you live in a warm climate, sow them directly outdoors. Otherwise, start them indoors in biodegradable pots that can be transplanted directly into the garden. Plant your seeds or seedlings in hills 4 to 6 inches high, 12 inches wide, and 4 to 8 feet apart.

2. When their stalks change from green to yellow or brown, your gourds are ready for picking. Use pruning shears to cut the gourds from the vine, leaving a 3- to 4-inch stem. Clean the gourds with a damp cloth and spread them on newspapers or wooden pallets in a dry, well-ventilated area. Turn them every few days to prevent rotting. The drying process can take several months. The outer skin will harden in a few weeks, but the inside will take much longer. As the water evaporates through the skin, a mold will naturally form on the outside of the gourds. Don't worry—this is fine!

3. When your gourds feel lightweight and the seeds inside rattle, it's time to turn them into birdhouses. Clean the gourds in warm, soapy water, using a scrub brush to remove the black, moldy outer layer. This may take a little elbow grease.

continues

gourd birdhouse, continued

4. Decide what species of bird you'd like to attract, and consult the box below for the entry hole dimensions for that particular species. Cut the entrance hole with a hole saw. Clean out the inside of the gourd by emptying the dried seeds and debris from the hole. Drill two small holes at the top of the gourd. Insert wire, twine, or a leather cord and loop it to form a hanger. Drill four ¼-inch holes in the bottom of the gourd for drainage. If you'd like to personalize your birdhouse, you can decorate it with paint or a wood-burning tool. Hang your gourd birdhouse in a sheltered location at the appropriate height (see box below).

TYPE OF BIRD	SIZE HOLE (IN)	HEIGHT ABOVE GROUND (FT)
bluebird	1½″	5-10
chickadee	1⅛″	6-15
nuthatch	1¼″	6-10
titmouse	1¼″	6-15
wren	1½″	6-10

wildflower seed paper

A garden isn't complete without wildflowers! Even if vegetables and herbs are your focus, incorporating colorful blooms will attract pollinators and add visual interest to the garden. We typically grow wildflowers directly from the seed packet, but we've also enjoyed making wildflower seed paper to plant or give away as gifts. One year we made paper in the shape of hearts and planted them in memory of our beloved dog Ruby. We've also given them as gifts for Valentine's Day, Mother's Day, and Earth Day. They are a treat to make and a creative way to upcycle used paper.

MATERIALS

3 cups shredded recyclable paper or newspaper
bowl
food processor or blender
fork
2 packets native wildflower seeds
strainer
wooden spoon
baking sheet or window screen
pencil
scissors

DIRECTIONS

1. Place the paper in a bowl, cover it with 3 cups warm water, and soak it for several hours.

2. Transfer the paper and water to a blender or food processor and pulse it until it has the consistency of pulp. Add more warm water, if needed.

3. Transfer the mixture back to the bowl, add the seeds, and mix well with a fork until the seeds are evenly dispersed. Pour the pulp into a strainer over the sink and press gently with the back of a spoon to strain out most of the water.

4. Transfer the pulp to a baking sheet or screen. Spread it out and press it flat with the palms of your hands until it's about ¼ inch thick. Allow the pulp to dry overnight. If it's still damp, carefully flip it to the other side and allow it to dry a second day.

5. Once the paper is dry, transfer it to a flat, dry surface. Draw shapes on it in pencil and carefully cut them out using scissors.

6. To use, place the cards in soil, cover them with a thin (⅛-inch) layer of compost, and water evenly every day until they sprout.

edible flower lollipops

These edible flower lollipops are surprisingly easy to make. They make wonderful gifts, party favors, or just a fun springtime or summertime treat. If you don't have access to fresh, organic flowers, you can order dried flowers and herbs online. Just be sure to double-check that your flowers are edible (see the list on page 55). Some of my favorites are borage, calendula, bush marigolds, chamomile, lavender, nasturtiums, pansies, roses, sunflowers, and violas. Yields 12 lollipops.

INGREDIENTS AND MATERIALS

lollipop mold
cooking spray or oil
edible flowers or flower petals (unsprayed)
saucepan
2 cups organic cane sugar
⅔ cup organic light corn syrup
⅔ cup water
confectionary thermometer
plant-based food coloring (optional)
lollipop sticks

DIRECTIONS

1. Coat the lollipop mold with a light coat of cooking spray or oil.

2. Place the flowers or petals facedown in the molds. Set aside.

3. In a small saucepan, combine the sugar, corn syrup, and water. Attach the confectionary thermometer to the saucepan and cook over medium heat until the mixture reaches the hard-crack stage at 302°F.

4. Remove the pan from the heat and stir in a small drop of food coloring (if using). As soon as the mixture stops bubbling, use a metal spoon to pour the syrup into the lollipop molds.

5. Quickly add the lollipop sticks and turn them a half rotation to ensure they're covered with syrup.

6. Allow the candy to cool and harden completely at room temperature, then remove the lollipops from the molds and enjoy!

NOTE: *To clean up, re-melt any remnants of the syrup left in the saucepan, then compost or toss it before it hardens again. To get the sticky residue off your pot, fill it with soapy water, bring to a boil, remove from the heat, and scrub the pan over the sink.*

summer bouquet ice lollies

What's better than lemonade in July? Not much! Unless the lemonade is frozen and infused with flowers. Come midsummer, I like to send my daughter and her friends out into the garden with a little basket and one simple chore: to collect a handful of bright edible flowers and herbs. Usually, they dawdle and get a little lost back there, but eventually they come back with sunflower petals, nasturtiums, pansies, thyme blossoms, pea blossoms, lemon balm, and mint—and a bit of sweat dripping from their brow. We add their harvest to homemade lemonade and freeze it to make these refreshing botanical pops. Yields 6 ice lollies.

INGREDIENTS AND MATERIALS

½ cup lemon juice
½ cup raw honey
3 cups filtered water
¼–⅓ cup cranberry juice (optional)
ice lolly molds
handful of edible flowers or herbs
ice lolly sticks (if your mold requires them)

DIRECTIONS

Mix the lemon juice and honey with the water until the honey dissolves. To make pink lemonade, add the cranberry juice, starting with ¼ cup and adding more until you're happy with the color. Pour the lemonade into six ice lolly molds, leaving ¼ inch of headspace. Wash the flowers or herbs and carefully add them, one at a time, to the molds. Add the cover and sticks to the molds and freeze for a few hours or until frozen.

hibiscus-mint sun tea

Here's a healthy version of Kool-Aid, made the slow, natural way. Instead of using boiling water to infuse the herbs, enlist the help of the sun! Set the tea out on a picnic table on a warm, sunny day, and watch the hibiscus petals turn the water a bright, cheery red. If you're lucky enough to grow hibiscus, you can use fresh flowers from your garden. Otherwise, you can purchase hibiscus tea at many grocery stores or online. Yields approximately 9 cups tea.

INGREDIENTS

2 cups fresh hibiscus flowers or ¼ cup dried hibiscus flowers
1 cup fresh peppermint leaves or ¼ cup dried peppermint leaf, plus more fresh leaves for garnish
1 tablespoon chopped lemon peel
¼ cup raw honey or more, to taste
2 cups ice cubes
lemon slices, for garnish

DIRECTIONS

Put the hibiscus, peppermint, and lemon peel in a quart-size mason jar and cover with 4 cups of water. Seal the jar with an airtight lid and let it steep outside, in the warm sunshine, for 1 hour. Pour the tea through a fine-mesh sieve into a large pitcher. Compost the herbs. Add the honey and stir to combine. Add 3 cups cold water, the ice cubes, and lemon slices, and serve with a sprig of mint.

lilac scones

We have a lot of false starts to spring here in the Midwest, but by the time the lilacs bloom you can rest assured that winter has passed. If that's not reason enough to love this beautiful plant, its intoxicating fragrance and many edible uses will probably do the trick. The easiest way to enjoy lilacs is by adding their flowers to a glass of ice-cold water. You can also infuse them in honey, add them to simple syrup, or incorporate them into cakes, pies, pancakes, and scones. Here's a classic recipe for scones that has become a springtime staple at my house. My family likes to top our scones with the dandelion syrup on the next page. Yields 8 scones.

INGREDIENTS

1 cup lilac flowers, stems removed
3 cups all-purpose flour
⅓ cup coconut sugar
2½ teaspoons baking powder
½ teaspoon baking soda
1 teaspoon salt
12 tablespoons cold unsalted butter
1 cup heavy cream
½ cup chopped toasted almonds
1 teaspoon vanilla extract

DIRECTIONS

1. Place the lilac flowers in a colander and rinse them with a gentle stream of cold water. Place them on a towel and allow them to air-dry.

2. Preheat the oven to 425°F. Line a rimmed baking sheet with parchment paper or a silicone baking mat.

3. In a large mixing bowl, combine the flour, sugar, baking powder, baking soda, and salt and mix well. Cut the butter into small cubes and use a pastry cutter or fork to work it into the dry ingredients. Continue until the butter looks like pea-sized crumbs. Add the cream, almonds, and vanilla and mix well. Fold the lilac flowers into the dough.

4. Turn the dough out onto a lightly floured surface. Work it into a ball, flatten it into a ¼-inch round disk, and cut it into 8 equal-size triangles. Transfer the triangles to the prepared baking sheet. Place the baking sheet in the freezer for 5 to 10 minutes until the dough is thoroughly chilled.

5. Bake for 12 to 15 minutes or until the edges of the scones are golden and crisp. Allow to cool and serve warm.

dandelion flower syrup

As adults, we're taught to believe dandelions are weeds, unless we're lucky enough to run into an herbalist who will quickly remind us that, not only are they edible, they're also nutritious! The flowers and leaves can be tossed into salads and used for making wines and honey, and the roots and leaves can be decocted or infused for tea.

Although dandelions can be enjoyed in a variety of recipes, my family loves making this syrup to sweeten pancakes, waffles, French toast, and scones. Just a word of caution: When foraging this plant, be sure to harvest away from roads, buildings, industrial sites, or any place where pesticides are sprayed. If that's not possible, purchase organic dandelion flowers online. Yields approximately 1½ cups of syrup.

INGREDIENTS

1½ cups unsprayed dandelion flowers
2½ cups organic cane sugar
½ cup honey
juice of 1 lemon

DIRECTIONS

1. Wash the flowers and dry them on a tea towel. With a knife, remove the petals as close to the base as possible.

2. Place the petals in a medium-size pot and cover them with 3 cups of water. Bring to a boil, then turn off the heat. Cover the pot with a lid and allow the tea to steep overnight.

3. The next morning, strain the tea into a bowl and use the back of a spoon to squeeze out as much liquid as possible. Compost the dandelions.

4. Return the tea to the pot. Add the sugar, honey, and lemon juice and bring to a simmer over low heat. Cook, stirring occasionally, for 1 to 1½ hours, or until the mixture thickens into a syrup.

5. Test the syrup's consistency by spooning a little onto a dish and cooling it in the refrigerator for a few minutes. If it is thinner than you would like, continue to cook the syrup for a few more minutes.

6. Let the syrup cool, then transfer it to an airtight glass container and store it in the refrigerator for up to 3 months.

sunflower house

When it comes to garden hideaways, the next best thing to building a bean teepee (page 199) is building a sunflower house. Similar to a bean teepee, the walls are made of plants except, instead of climbing vines, they're made from strong, tall sunflowers. When planted in a square, rectangular, or circular formation, these bright and lofty flowers create the perfect hiding spot or reading nook. Just be sure to select a tall sunflower variety, such as Mammoth or California Greystripe, that will grow to at least 6 feet in height.

MATERIALS

rope, hose, or stakes and twine
shovel and garden hoe
weed-free straw or mulch (optional)
sunflower seeds
compost
window screening

DIRECTIONS

1. Choose a location for your house that is flat and gets 6 to 8 hours of direct sunlight a day. Choose a shape for your house and mark it with a rope, hose, or stakes and twine. Leave an opening for the doorway that is approximately 2 feet wide.

2. If you wish to make it weed-free, remove any rocks, grass, and weeds from the inside area. Otherwise, simply trim vegetation as needed during the growing season.

3. Using the string (or rope or hose) around the perimeter as a guide, use a hoe to clear the vegetation from the house's perimeter. With a shovel, dig a 3-inch-deep trench, 6 to 12 inches wide.

4. Poke a 1-inch-deep hole every 6 inches along the perimeter of your house. Place two seeds in each hole and cover them with compost. You can plant a single row, or plant double rows if you want thicker walls. Water the seeds well.

5. To protect the seeds from birds, cover them with window screen material held down with rocks. Water the seeds daily, keeping them consistently moist, and remove the window screen material once the seeds sprout. Mulch with straw or other organic mulch if desired.

6. Continue to water the young seedlings well. As the plants mature, their roots will reach deeper into the ground to access water below the surface, at which point you can water them less frequently.

savory herbal biscuits

There aren't many things my daughter loves more than baking. Since the time she was tall enough to stand on a chair and open the kitchen cabinets, she's been mixing flour, sugar, and butter to make all sorts of delicious treats. To this day, she bakes to self-soothe the way I garden to calm my nerves. Sometimes I'm able to fuse our two passions by baking with herbs and flowers straight from our garden.

These savory herbal biscuits are one of our favorite garden recipes. I like to fold rosemary leaves into the dough and decorate the tops with more delicate herbs, like coriander, lavender, parsley, sage, or thyme. Although these are an obvious summer snack, they're also the perfect biscuit to serve at holiday parties or to gift to friends and family. Yields 18 to 24 biscuits, depending on what size cookie cutters you use.

INGREDIENTS

1¼ cups all-purpose flour, plus more for dusting

1 cup grated Parmesan cheese

2 tablespoons chopped fresh herb leaves, plus more whole leaves for decorating (optional)

½ teaspoon salt

½ cup unsalted butter, cubed, at room temperature

DIRECTIONS

1. Line a rimmed baking sheet with parchment paper and preheat the oven to 350°F.

2. Place the flour, Parmesan, chopped herbs, and salt in the bowl of a food processor and pulse to combine. Add the butter and pulse for about 1 minute, or until the ingredients come together and form a dough. If they don't form a dough, add 1 to 2 tablespoons water.

3. Move the dough to a lightly floured surface and knead it for a few minutes until it's nice and smooth.

4. Roll out the dough to about ¼-inch thickness. If the dough sticks to your rolling pin, lightly flour the dough, set it between two sheets of parchment paper, and roll it out that way.

continues

savory herbal biscuits, continued

5. Cut the biscuits using a round cookie cutter or the mouth of a half-pint-size mason jar. Use a spatula to carefully transfer the biscuits to the baking sheet. If there's extra dough, re-roll it and cut more biscuits.

6. If you want to garnish the biscuits with fresh herbs, lay the herbs on top of each biscuit, place a piece of parchment paper over the top, and gently roll the rolling pin over the biscuits to press the herbs into the dough. Or just sprinkle herbs on top.

7. Bake for 10 to 12 minutes, or until the biscuits turn slightly golden on the bottom. Watch them closely, as they brown quickly!

8. Remove the biscuits from the oven and allow them to cool for 10 to 15 minutes.

9. Serve immediately or freeze in an airtight container for up to 4 weeks.

botanical shortbread cookies

My children and I enjoy making these shortbread cookies using whichever edible flowers are in season. In the spring, we decorate them with pansies and violas, which work particularly well because of their thin, delicate blossoms. Come summer, we use the petals of geraniums, lilacs, marigolds, nasturtiums, roses, and sunflowers. Yields 18 to 24 cookies.

INGREDIENTS

2 dozen edible flowers or 2 cups edible flower petals
1 cup butter, softened
½ cup confectioners' sugar
2 cups all-purpose flour, plus more for dusting

DIRECTIONS

1. Snip the flowers from their stems, being careful to keep them intact. If the flowers are too big for the cookies, remove the petals and use them individually.

2. Press the flowers by placing them between two sheets of parchment paper and setting the paper between heavy books for at least an hour. If you have a flower press, you can use that instead.

3. When you're ready to bake, preheat the oven to 325°F. Line two baking sheets with parchment paper and set aside.

4. Place the butter, confectioners' sugar and flour in the bowl of a stand mixer and mix well until a crumbly dough forms. (It will look like crumbs in the bowl, but will form a ball when you squeeze the dough in your hand.)

5. Dust a work surface with flour. Roll the dough into a ball and flatten it with a rolling pin to about ¼-inch thickness. If the dough is sticky, lightly flour it and roll it out between two sheets of parchment paper. Cut out cookies using cookie cutters or a half-pint-size mason jar. Use a spatula to transfer the cookies to the baking sheets.

6. Bake for 20 minutes, or until the cookies turn slightly golden on the bottom.

7. Remove the cookies from the oven. While they're still hot, use your fingers to gently press the flowers onto the tops of the cookies.

8. Allow the cookies to cool, then serve or freeze in an airtight container for up to 4 weeks.

herbal soup wreaths

Here's a delightful way to use and share herbs from your garden. Just braid trimmings into pint-size wreaths and toss them into broths and brines. After they flavor your soups or stews, simply fish them out and toss them in the compost bucket. There's more versatility to making them than you might imagine. Enjoy them fresh or dried, or pop them in the freezer and preserve them for later. Make them with one herb, two herbs, three, or four. Just keep in mind that an herb's flavor intensifies as it dries. Potent herbs like oregano and sage can pack a punch if you include too much of them in one wreath.

MATERIALS

scissors
fresh culinary herbs
natural thread or twine
clipboard (optional)

DIRECTIONS

1. Using scissors, snip fresh herb sprigs 6 to 10 inches long. Remember that you'll be braiding them with other herbs, so choose flexible stems when possible.

2. Wash the herbs and allow them to dry thoroughly to prevent them from molding on the wreath.

3. Choose three strands of herbs to braid. Bind the pieces together on one end with thread or twine.

4. If you have a clipboard, place the tied end of the herbs under the clamp to hold your herbs in place; otherwise, have someone hold the end steady or braid freehand. Braid the herbs together until you reach the other end.

5. Join the two ends to form a loop (or wreath) and tie them together with thread or twine. You can also try using another herb, such as chives, to tie the ends together.

6. Use the wreaths fresh, hang them to dry, or freeze for later.

NOTE: *If you plan to add your wreaths to soups or stews (as opposed to using them for decoration), be sure to tie them up with a natural thread that is safe for cooking. Avoid synthetic fibers such as nylon or polyester and opt instead for organic hemp, jute, or cotton.*

gardener's hand scrub

This hand scrub is a wonderful gift for fellow gardeners and an easy craft for kids or adults. The ingredients remove dirt and grime and leave hands feeling soft and clean. Yields 1½ cups.

INGREDIENTS

1 cup sugar (white, brown, or coconut)
¼ cup coconut oil, melted
¼ cup castile soap
zest of 1 lemon
1 tablespoon fresh rosemary leaves (stems removed)
10 drops lemon essential oil
5 drops rosemary essential oil

DIRECTIONS

In a medium bowl, combine the sugar, coconut oil, castile soap, lemon zest, and rosemary leaves. Mix well to create a paste. Add the essential oils and stir well. Transfer the scrub to a clean 12-ounce glass jar, seal tightly with a lid, and keep by the sink (or package it to give as a gift). If temperatures rise above 76°F, the coconut oil may melt, in which case you can store the scrub in the refrigerator. On the counter or in the refrigerator, this scrub keeps for about 2 months.

herb and spice fire starters

Whether you're camping outdoors or sitting at home on a chilly night, nothing beats cozying up to a warm fire. We like to jump-start our fires with the help of fire starters made from aromatic herbs and fragrant spices. For the wax, I prefer beeswax, but soy wax works well too. You can also melt down and repurpose old candle stubs—just toss them in a double boiler and carefully remove the old wicks once the wax melts. Yields 12 fire starters.

MATERIALS

12-cup standard muffin pan
12 standard baking cups
12 bay leaves
6 cinnamon sticks, broken into small pieces
12, 1-inch sprigs rosemary
12 mini pine cones (optional)
about 3 teaspoons dried herbs
about 3 teaspoons dried spices
12, 1½-inch pieces cotton candle wick
double boiler
4 cups natural beeswax flakes, soy wax flakes, or old candle stubs
wooden chopstick

DIRECTIONS

1. Line the muffin pan with the baking cups. Fill each baking cup with 1 bay leaf, a few pieces of broken cinnamon sticks, 1 sprig of rosemary, 1 mini pine cone, ¼ teaspoon dried herbs, and ¼ teaspoon dried spices. Place a cotton wick in the center of each cup.

2. In the double boiler, melt the wax flakes. Pour melted wax into each cup, filling them three-quarters of the way. If needed, use the chopstick to help submerge the ingredients and the wick into the wax. Set the cups aside to cool.

3. Once the wax starters have solidified, pop them out of the baking cups. Store them in an airtight container or package them individually for gifts.

4. To use a fire starter, place it at the bottom of your firepit or at the bottom center of a wood-burning fireplace, and light the wick. Do not use in a gas fireplace or wood-burning stove. Each firestarter should burn for approximately 20 minutes.

citronella lemon bowl candles

These candles are a great way to enjoy evenings outdoors without having to share the garden with mosquitoes. Children can help make them by cutting the lemons, scooping out their flesh, and sprinkling the candles with fresh herbs. To ensure the candles repel mosquitoes, infuse and garnish them with herbs and essential oils such as citronella, eucalyptus, lavender, and rosemary. As with the fire starters on page 226, you can use beeswax or soy wax for this recipe. Yields 8 lemon bowls.

MATERIALS

4 lemons
double boiler
20 ounces beeswax or soy wax pellets
8 drops citronella essential oil
8 drops eucalyptus essential oil
wooden craft stick
8, 1½-inch cotton wicks
8 wooden clothes pegs
dried or fresh rosemary and lavender

DIRECTIONS

1. Slice the lemons in half and hollow out the flesh with a knife and a spoon.

2. Pour 2 inches of water in the bottom of the double boiler and bring it to a simmer over medium heat.

3. Add the wax pellets to the top of the double boiler and heat them until they melt.

4. Remove the top of the double boiler from the heat and allow the wax to cool for 5 minutes. Add the essential oils and mix thoroughly with the craft stick.

5. Thread the top of each wick through the hole in the center of a clothes peg and lower the clothes peg onto the rim of the lemon bowl. The clothes peg will hold the wick in place while you pour the wax and until the wax cools and solidifies. Repeat with the remaining lemon halves.

6. Pour the melted wax mixture into each lemon bowl. Re-center the wicks if needed and sprinkle the top of each candle with rosemary and lavender.

7. Allow the candles to cool, then use them within 2 days, before the lemons wilt.

PLANT PROFILES

Although you can surmise a lot about a plant by knowing what family it belongs to, information specific to each type of plant is invaluable in helping you decide when, how, and what to grow. In this section, you'll find profiles for the most common garden crops, including how to start, grow, and harvest them, as well as potential problems you may encounter along the way. Spacing suggestions are given for each plant, although keep in mind that they apply to raised row gardens; if you're growing in containers or raised beds, plants can be planted a bit more densely.

This section is meant to serve as a reference, but it doesn't cover everything. One crop may have dozens, if not hundreds, of different cultivars bred for different characteristics and with different growing needs. If the information on a seed packet or plant tag says something different than what you find here, by all means follow the instructions that came with your plant. Hang on to the seed packet for future reference, and keep notes in your gardening journal or in this book.

Becoming a great gardener is all about hands-on learning. A book is a great tool for getting started and for at-a-glance answers, but there's no better teacher than experience.

For convenience, plants are organized in alphabetical order, rather than by season. For planning purposes, refer to the box below to figure out what to grow when.

COOL SEASON		WARM SEASON		PERENNIAL HERBS
asparagus	leeks	aubergine	peppers	chives
beetroot	lettuce	basil	strawberries	mint
broccoli	onions	corn	summer squash	rosemary
Brussels sprouts	parsley	cucumbers	sweet potatoes	sage
cabbage	peas	green beans	tomatoes	thyme
carrots	potatoes	melons	winter squash	
coriander	radishes			
dill	rocket			
garlic	spinach			
kale	Swiss chard			

asparagus

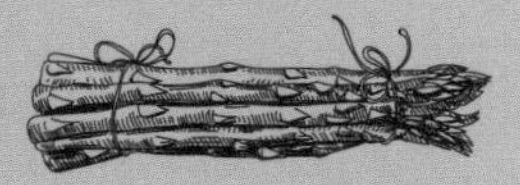

FAMILY: Asparagaceae
GROWING SEASON: perennial
PLANTING: start indoors or direct-sow outdoors
START INDOORS: 3 months before the last spring frost
SOW OUTDOORS OR TRANSPLANT: after the last spring frost
SUN NEEDS: 6 to 8 hours per day
WATER NEEDS: high
ROTATION: avoid following onion family crops
ZONES: 2–9
PLANTS PER SQUARE FOOT: 1

OVERVIEW: Asparagus is a perennial that can take 2 to 3 years to produce but will come back annually for 15 to 30 years once established. The edible part of the asparagus plant is the young stem shoot, which emerges in spring as soil temperatures rise above 50°F. It produces one of the earliest and tastiest spring vegetables.

START: Although asparagus can be grown from seed, most gardeners start with 1-year-old crowns, which can be planted after the last spring frost, as soon as the soil can be worked. To plant crowns, dig a trench about 12 inches deep and 12 inches wide, then plant crowns 18 inches apart. If planting more than one row, space the trenches at least 3 feet apart. Seeds should be started indoors 3 months before the last spring frost. Soak the seeds in water for up to 24 hours before sowing them ¾ inch deep in 4-inch pots. Once plants reach 12 inches in height, harden them off and transplant them 6 to 8 inches deep in a temporary site in the garden. When the plants mature in the autumn, identify the berry-less male asparagus plants (which are more productive than their female counterparts) and transplant them to a permanent site.

GROW in full sun and in fertile, well-draining soil with a pH of at least 7.0. Add lime or wood ash to adjust the pH if necessary. Because asparagus is a perennial, it needs to produce a lot of energy—enough to produce shoots, survive winter, and make new ferns. Side-dress with a layer of compost and well-aged manure in the spring and autumn to ensure plants have plenty of nutrients. Weed well and often, as asparagus does not like to share nutrients with other plants.

HARVEST: For the most productive asparagus beds, wait to harvest until the third year after planting. Harvest spears for up to 6 weeks in early spring. In subsequent years, harvest for up to 8 weeks. To harvest, when spears are no more than 7 inches tall, cut them with a sharp knife, 1 inch below the soil line. In warm weather, harvest every 2 to 3 days for the best quality.

CONTAINERS: Asparagus grows well in containers, although plants won't produce or last as long as garden-grown crops. Choose a container that is at least 20 inches deep and 20 inches wide.

PROBLEMS: The biggest concern for asparagus plants is weeds. Hand-pull on a regular basis in spring and early summer. Applying a mulch of shredded leaves or straw can also help with weed control.

aubergines

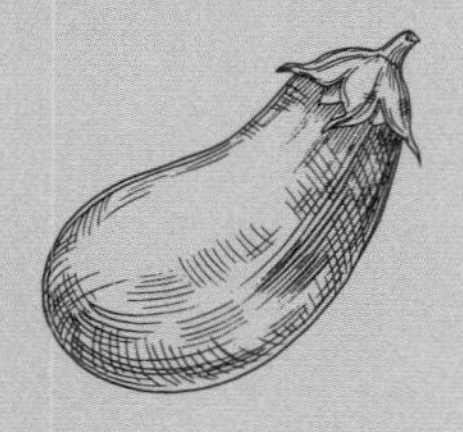

FAMILY: Solanaceae
GROWING SEASON: warm season
PLANTING: start indoors or purchase transplants
START INDOORS: 6 to 8 weeks before the last spring frost
TRANSPLANT: 2 weeks after the last spring frost
SUN NEEDS: 6 to 8 hours per day
WATER NEEDS: moderate
ROTATION: plant after beans or peas
PLANTS PER SQUARE FOOT: 1

OVERVIEW: Aubergines grow wild as a perennial in their native homeland of South Asia, but most gardeners in other parts of the world grow them as an annual. Because of their tropical and subtropical nature, aubergines require warm temperatures and can be difficult to grow in cold climates.

START: Aubergines have a long growing season and take 100 to 120 days to grow from seed to harvest. To get a jump-start on the growing season start seeds indoors or purchase transplants from a nursery. If possible, plant seedlings in raised beds or containers, where the soil tends to warm up faster. Alternatively, spread a piece of black plastic over the soil to warm it before planting. Sow seeds indoors, ¼ inch deep, in 4-inch pots and keep them warm with a heating mat or by placing the seed trays on top of a warm refrigerator. One week before transplanting seedlings, add compost or aged manure to the soil in the garden bed. When daytime temperatures are consistently in the 70°F range, transplant seedlings 12 to 24 inches apart. Be sure to set up support structures, such as stakes or cages, at the time of planting.

GROW aubergines in a sunny location with rich, well-draining soil with plenty of organic matter. Add compost or aged manure to nourish plants and mulch to heat the soil and conserve moisture. In cold climates, consider using row covers to keep plants warm, opening the sides on warm days to allow for pollination.

HARVEST: Aubergines should be picked as soon as they show signs of being ripe, as slightly immature fruits taste best. Knowing when to harvest aubergines is a bit of an art. The skin of the aubergine should be smooth, glossy, and unwrinkled. If you squeeze the aubergine, it should spring back immediately. If indentations remain in the fruit, it's not ripe yet. To harvest an aubergine, use a sharp knife to cut the fruit from the stem, leaving about 1 inch of the stem attached. Aubergines deteriorate rapidly, so don't wait too long to eat them once harvested.

CONTAINERS: A single plant will grow well in a container that is at least 12 to 14 inches in diameter and 12 inches deep.

PROBLEMS: Aubergines are prone to a few pests and diseases, including aphids, caterpillars, mealybugs, red spider mites, whitefly, blossom-end rot, and botrytis.

basil

FAMILY: Lamiaceae
GROWING SEASON: warm season
PLANTING: start indoors or direct-sow outdoors
START INDOORS: 6 weeks before the last spring frost
SOW OUTDOORS OR TRANSPLANT: after the last spring frost
SUN NEEDS: 6 to 8 hours per day
WATER NEEDS: moderate
ROTATION: n/a
PLANTS PER SQUARE FOOT: 2

OVERVIEW: Basil is an annual warm-season herb that tastes great in many dishes and is perhaps best known for its use in pesto. The most common type of basil is sweet basil; other varieties include lemon basil, purple basil, and Thai basil. Basil is also valued for its medicinal uses. It acts primarily on the digestive and nervous systems, easing gas and stomach cramps and preventing and relieving nausea and vomiting. Basil is also antibacterial, and the juice of its fresh leaves can be used to relieve the itch and pain of an insect bite or sting.

START SEEDS indoors 6 weeks before the last spring frost or direct-sow outdoors once the soil has warmed to at least 50°F. Plant seeds ¼ inch deep, 10 to 12 inches apart.

GROW in full sun and in fertile soil that is moist but well draining. Basil makes the ideal companion plant to tomatoes and is believed to increase the flavor and color of its fruit and repel aphids and other insect pests.

HARVEST: Begin harvesting basil shortly after it sprouts, once the second set of leaves appear. Harvest leaves one to two at a time at first, then begin snipping the stems above the node where two leaves meet. Harvest regularly to encourage plants to grow full and bushy. If flowers begin to appear, pinch or snip them off so plants can focus their energy on producing tasty leaves instead.

CONTAINERS: Basil grows well indoors or outdoors in containers that are at least 4 inches deep.

PROBLEMS: Basil is prone to a few pests and diseases, including aphids, white powdery mildew, and a variety of fungal and bacterial diseases, such as Fusarium wilt. Seedlings may also experience damping off. Avoid problems with diseases by waiting until the weather warms up before planting outside and taking care not to overcrowd plants.

beetroot

FAMILY: Amaranthaceae
GROWING SEASON: cool season
PLANTING: start indoors or direct-sow outdoors
START INDOORS: 6 to 8 weeks before the last spring frost
SOW OUTDOORS OR TRANSPLANT: 2 to 3 weeks before the last spring frost
SUN NEEDS: 4 to 6 hours per day
WATER NEEDS: moderate
ROTATION: avoid following spinach or Swiss chard
PLANTS PER SQUARE FOOT: 9 (large varieties); 16 (small varieties)

OVERVIEW: Beetroot is a biennial plant that is typically grown as an annual. It is also one of the easiest vegetables to grow. Not only is it less susceptible than most veggies to pests and diseases, but it also doesn't need staking, pruning, or fussing of any kind. Beetroot takes up little space and is grown for its nutrient-dense leaves as much as for its colorful roots. There are two main types of beetroot: globe-shaped and long-rooted. Both come in a variety of shades, such as red, orange, gold, yellow, and white.

START: Beetroot can be started indoors or outdoors. If starting it outdoors, sow seeds 2 to 3 weeks before the last frost date, when soil temperatures are at least 50°F. Plant seeds ½ inch deep and 1 inch apart, then thin seedlings to 3 to 4 inches apart. Beetroot seeds have a better germination rate if you soak them for 24 hours before sowing. It also helps to keep the seeds evenly moist until they sprout.

Each beetroot seed is actually a cluster of 2 to 6 seeds inside a dried fruit, so you will need to thin seedlings to 3 to 4 inches apart once the greens are about 4 inches tall. Rather than pulling the seedlings up by their roots, cut them at soil level. For a continuous harvest, sow seeds every 2 to 3 weeks. For an autumn harvest, sow seeds from late summer through early autumn, starting about 4 to 6 weeks before the first autumn frost.

GROW in full sun or partial shade in rich, well-draining soil. Because beetroot grows best in cool and moist conditions, it helps to have a bit of shade and to mulch and water it regularly. Excess nitrogen can result in lush tops and tiny roots; if you fertilize beetroot, be sure to use a supplement that is low in nitrogen and high in phosphorus and potassium.

HARVEST beetroot between 55 and 70 days after planting, when the roots are between the size of a golf ball and a tennis ball (larger roots tend to be tough and woody). Pick beetroot greens any time, being sure to leave a few behind so the plant can continue to grow.

CONTAINERS: Beetroot grows well in containers at least 10 inches deep. Small, globe-shaped varieties do particularly well in containers.

PROBLEMS: The most common problem with Beetroot is that it ends up being too small or malformed as a result of too much shade or overcrowding.

broccoli

FAMILY: Brassicaceae
GROWING SEASON: cool season
PLANTING: start indoors or direct-sow outdoors
START INDOORS: 9 weeks before the last spring frost for a spring crop
TRANSPLANT: 3 weeks before the last spring frost
SOW OUTDOORS: in early summer for an autumn crop
SUN NEEDS: 4 to 6 hours per day
WATER NEEDS: moderate
ROTATION: avoid following other crops in the brassica family
PLANTS PER SQUARE FOOT: 1

OVERVIEW: Broccoli is a hardy biennial that is typically grown as an annual crop. Broccoli will bolt and go to seed in warm weather, so the biggest challenge to growing it is getting the timing right. Broccoli should be planted so it matures in spring, autumn, or early winter, when daytime temperatures average below 75°F. Broccoli is ready to harvest in 55 to 85 days when grown from transplants and 70 to 100 days when grown from seed. Broccoli is frost hardy and can withstand temperatures as low as 20°F.

START seeds indoors 9 weeks before the last spring frost. Transplant seedlings into the garden 3 weeks before the last spring frost. Plant seedlings 1 inch deeper than they grew in their pots, 12 to 18 inches apart. If planting broccoli as an autumn crop, direct-sow seeds in the garden ½ inch deep and 3 inches apart. Once the seedlings are 2 to 3 inches tall, thin them to 12 inches apart.

GROW in compost-rich, well-draining soil with a pH between 6 and 6.8. Keep the soil consistently moist, watering 1 to 1½ inches per week. Water at the base (not the leaves) to avoid rot, and side-dress plants with compost at the time of planting and again midseason.

HARVEST: Use a sharp knife to remove the central flower head along with 5 to 6 inches of stem. Leave the base of the plant and some outer leaves to encourage new heads or secondary shoots to develop.

CONTAINERS: Single broccoli plants grow well in containers that are at least 8 inches wide. Multiple plants can be grown in larger containers that allow for plants to be spaced 18 inches apart.

PROBLEMS: Broccoli is susceptible to cutworms, and cabbage caterpillars. Use row covers to prevent pests, and pick them off by hand or spray with *Bacillus thuringiensis* (Bt). Broccoli is prone to diseases common to plants in the cabbage family, including clubroot and downy mildew. To avoid these problems, plant disease-resistant varieties, rotate crops each year, and remove infected plants immediately.

brussels sprouts

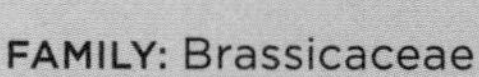

FAMILY: Brassicaceae
GROWING SEASON: cool season
PLANTING: start indoors or direct-sow outdoors
START INDOORS: 9 weeks before the last spring frost
TRANSPLANT: 6 weeks before the last spring frost
SOW OUTDOORS: 4 months before the first autumn frost
SUN NEEDS: 6 to 8 hours per day
WATER NEEDS: high
ROTATION: avoid following other crops in the brassica family
PLANTS PER SQUARE FOOT: 1

OVERVIEW: Brussels sprouts require a long growing season of 80 to 100 days and do best when grown as an autumn or early winter crop. As one of the hardiest crops in the garden, they prefer a bit of a nip in the air and taste best after exposure to frost.

START: To determine the best planting time for your area, count backwards from your first autumn frost date and direct-sow seeds in the garden 4 months before that date. Plant seeds ½ inch deep and 2 to 3 inches apart, then thin seedlings to 12 to 24 inches apart. Plant transplants 12 to 24 inches apart.

GROW in full sun in cool, evenly moist soil. Feed plants with a nitrogen-rich organic fertilizer twice a season—once when the plants are about 12 inches high and again 4 weeks later. Mulch the soil both to retain moisture and keep soil temperatures cool. Prune leaves that turn yellow so plants can put their energy into new growth. If plants become top-heavy, stake and tie them with a 2-foot long wooden stake 2 inches away from the base of the plant.

HARVEST sprouts from bottom to top when they are about 1 inch in diameter or before they start to open. To force sprouts to mature all at once, remove the top of the stalk when the lowest sprouts are about ½ to 1 inch in diameter—the entire stalk of sprouts should mature within 2 weeks. To remove sprouts, snap or cut them off from the stem. If the sprouts haven't fully matured before temperatures dip below 25°F, cut the entire plant at the base, strip the leaves, and hang the stalk upside down in a cool, dark place for up to 3 weeks or until the sprouts are ready to be picked.

CONTAINERS: Single plants can be grown in a 5- to 7-gallon container, at least 12 inches deep and 12 to 14 inches in diameter.

PROBLEMS: Brussels sprouts are not difficult to grow, but they are heavy feeders and need fertile soil. They also take a long time to mature and require good planning to get the timing right. Brussels sprouts are affected by the same insects that plague cabbage plants, including aphids, earwigs, cutworms, leaf miners, nematodes, snails, and slugs.

cabbage

FAMILY: Brassicaceae
GROWING SEASON: cool season
PLANTING: start indoors or direct-sow outdoors
START INDOORS: 9 weeks before the last spring frost
SOW OUTDOORS OR TRANSPLANT: 3 weeks before last spring frost
SUN NEEDS: 3 to 4 hours per day
WATER NEEDS: moderate
ROTATION: avoid following other crops in the brassica family
PLANTS PER SQUARE FOOT: 1

OVERVIEW: Cabbage is a hardy, leafy vegetable that can be challenging to grow. Not only is it sensitive to warm temperatures, but it also attracts a variety of different pests. As a heavy feeder, it depletes the soil quickly and requires more attention to soil nutrition than most plants. It's worth learning to grow, though, if only because it adds so much character and charm to the garden.

One of the tricks to growing cabbage is getting the timing right. As a cool-weather crop, it can be started in the spring so it comes to harvest before the summer heat sets in or in mid- to late summer so it matures during the cool days of autumn or winter.

START: If starting seeds indoors, sow them ¼ inch deep, 9 weeks before the last spring frost. Transplant seedlings on a cloudy afternoon 3 weeks before the last spring frost, spacing plants 12 to 24 inches apart. For an autumn harvest, direct-sow seeds outdoors or transplant seedlings in mid- to late summer. If you live in a hot, dry climate, wait until late summer to start your autumn crop.

GROW in full sun in rich, well-draining soil with a pH of 6.0 to 7.5. Prepare the soil in advance by adding layers of aged manure or compost. Mulch well around the base of each plant to maintain soil moisture and moderate soil temperature. Apply a balanced fertilizer a few times during the growing season, especially if growth starts to lag. Practice crop rotation to prevent buildup of soilborne diseases.

HARVEST cabbage when the heads are firm and large by cutting each head at the base with a sharp knife. Remove any yellow leaves and store the head in a cool, dark place. To get a second crop, remove the cabbage head but leave the outer leaves, stems, and roots in the soil. The plant will produce multiple smaller heads that can be harvested when they're the size of tennis balls.

CONTAINERS: Dwarf varieties can be grown in pots that are at least 10 inches deep.

PROBLEMS: Cabbage can be a magnet to insect pests. It also can be difficult to grow, especially as a result of planting at the wrong time; under- or overwatering; letting weeds compete for nutrients; failing to add organic matter and fertilizer to the soil; or planting it in beds where broccoli, Brussels sprouts, cabbage, cauliflower, or kale were grown in the last two 2 years.

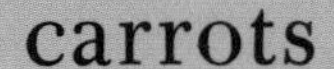

carrots

FAMILY: Apiaceae
GROWING SEASON: cool season
PLANTING: direct-sow outdoors
SOW OUTDOORS: 3 to 5 weeks before the last spring frost, and/or 4 to 8 weeks before the first autumn frost
SUN NEEDS: 4 to 6 hours per day
WATER NEEDS: moderate
ROTATION: avoid following celery, dill, fennel, parsley, or parsnips
PLANTS PER SQUARE FOOT: 16

OVERVIEW: Carrots are a biennial crop that is typically grown as an annual. They're fairly easy to grow as long as they're planted in loose, stone-free soil and in cooler weather. Beds improved with well-rotted compost are ideal for growing carrots; manure and fertilizer should be avoided, however, as they can cause carrots to fork. Although carrots are grown primarily for their roots, their leaves are perfectly edible and can be used to make a delicious pesto.

START: Carrots do not transplant well, so start seeds by direct-sowing them in the garden when danger of frost has passed. For a continuous harvest, succession-sow seeds every 3 weeks from early spring to midsummer so you can harvest roots from late spring to early winter. Before sowing, remove all stones from the soil, then rake it to a fine tilth. If the soil is rocky or compact, add several inches of compost or plant in containers or raised beds. Plant seeds ¼ inch deep, 3 inches apart. Cover seeds with a fine layer of compost and keep soil consistently moist during germination and growth.

GROW: Crowded seedlings should be thinned out carefully to prevent damaging neighboring plants. Rather than pulling weaklings out by the root, use scissors to snip off their foliage at the soil line. Keep the soil moist in dry weather to prevent carrots from bolting before they're mature. Regularly remove weeds to eliminate competition for nutrients and mulch to prevent the tops of the roots from turning green.

HARVEST: Most carrots are ready to harvest within 70 to 100 days. Although the days-to-maturity information on the back of a seed packet can help you determine when to harvest your carrots, you can also use your powers of observation to decide the best time to pick them. Most gardeners harvest their carrots when the shoulders are about ½ inch to ¾ inch in diameter. It's usually easy to measure the top because it's often poking out of the soil. If not, brush away the soil from around the top of the root to get a better look. To harvest carrots, either pull them up gently by their tops or dig them out of the soil with a trowel.

CONTAINERS: A short variety of carrot can be grown in a container at least 10 inches deep.

PROBLEMS: The carrot rust fly is the most common problem with carrots. Since these pests have a maximum flying altitude of around 12 inches, growing in raised beds is a good way to prevent them. Companion planting carrots with garlic, chives, or leeks will also help drive them away.

chives

FAMILY: Alliaceae
GROWING SEASON: perennial
PLANTING: start indoors or direct-sow outdoors
START INDOORS: 6 to 8 weeks before the last spring frost
SOW OUTDOORS OR TRANSPLANT: after the last spring frost
SUN NEEDS: 3 to 4 hours per day
WATER NEEDS: moderate
ROTATION: n/a
ZONES: 3–10
PLANTS PER SQUARE FOOT: 1

OVERVIEW: Chives are a cold-hardy perennial herb in the onion family that produces green leaves and edible blossoms. A cool-season crop, they grow best in the spring and autumn and will go dormant during the warmer months of summer. Chives make a great companion plant in the garden, as they deter pests with their strong odor and attract beneficial insects with their flowers.

START: To get a head start in cold climates, start seeds indoors 6 to 8 weeks before the last spring frost. Otherwise, sow seeds directly in the garden when the soil is workable. Sow seeds ¼ inch deep and 1 inch apart, and thin seedlings to 3 inches apart. Note that seeds are slow to germinate, so don't panic if you don't see sprouts right away! Propagate existing plants by dividing them in the autumn. To divide a plant, cut its leaves down to 3 inches above the base, dig up the plant, pull the roots apart into three or four smaller clumps, and replant them in new locations.

GROW in full sun in moist, rich, well-draining soil with a pH of 6 to 7. Before planting, add well-composted organic matter to the soil, as well as a layer of mulch to help retain soil moisture and suppress weeds. Chives are drought tolerant but will grow best when watered consistently. To increase yields, moisten the soil thoroughly when watering and use a nitrogen-rich organic fertilizer in late spring or early summer. Chives grow prolifically and can easily take over a garden if you allow their flowers to develop fully. To prevent them from spreading their seeds, be sure to remove their flowers when they bloom.

HARVEST: Chives can be harvested at any time once they're established. To harvest leaves, snip them with a pair of scissors 2 inches above the base. Begin harvesting from the outside leaves inward, three or four times the first year and once a month in subsequent years. Come winter, let chives die back (meaning its leaves die but its roots remain alive) or pot them and bring them indoors until spring.

CONTAINERS: Chives grow well in containers that are at least 4 inches deep.

PROBLEMS: Chives are fairly low maintenance. The most common pests and diseases to look for are damping off, downy mildew, pink root, onion maggots, and thrips.

coriander

FAMILY: Apiaceae
GROWING SEASON: cool season
PLANTING: direct-sow outdoors
SOW OUTDOORS: after the last spring frost
SUN NEEDS: 3 to 4 hours per day
WATER NEEDS: moderate
ROTATION: avoid following with other crops in the carrot family
PLANTS PER SQUARE FOOT: 9

OVERVIEW: Coriander is a fast-growing, aromatic annual herb that grows best in cool weather. You'll need to watch plants carefully as the weather gets warmer. Coriander has a short life cycle and bolts quickly when temperatures exceed 85°F.

START: Coriander is difficult to transplant, so it's best to sow seeds directly in the garden. Fortunately, it grows quickly, so there's really no need to start it indoors. Sow seeds ½ inch deep, 2 inches apart, then thin them to 6 to 12 inches apart once the seedlings are a few inches tall. For a continuous harvest, plant seeds in succession every 2 to 3 weeks until late summer.

GROW in moderately rich, well-draining soil with a pH of 6.6. If you live in a warm climate, mulch the soil to keep it evenly moist. Water and fertilize sparingly. To maximize foliage, pinch back young coriander plants and, unless you are growing it for seed, cut off the top part of the main stem as soon as it looks like it might be developing flower buds.

HARVEST: Coriander can be harvested any time after the plant is 6 to 8 inches tall. Young, immature leaves have the best flavor. Cut off just the top 2 to 3 inches to ensure continuous growth. Continue harvesting leaves until the plant flowers.

To harvest coriander seeds, allow the plant to flower and go to seed. When the seed heads turn yellowish brown, snip them and place them upside down in a brown paper bag. Close the bag and hang it in a cool, dry place while the seed heads continue to ripen. After a few weeks, the seed head will drop the seeds, at which time you can collect and store them in an airtight container.

CONTAINERS: Coriander grows well in containers at least 4 inches deep.

PROBLEMS: Coriander bolts quickly in warm weather and should be harvested before it gets too hot. It rarely has serious problems with insects and is often used as an insect repellent. The two diseases that most often affect coriander are leaf spot and powdery mildew.

corn

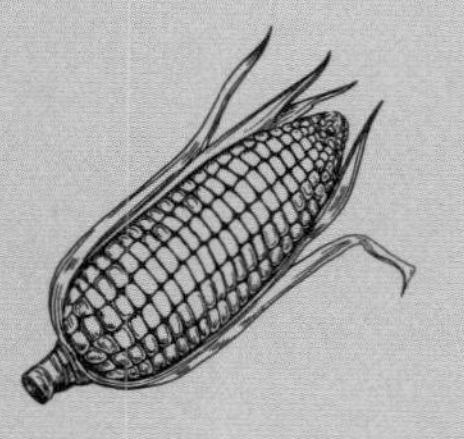

FAMILY: Poaceae
GROWING SEASON: warm season
PLANTING: direct-sow outdoors
SOW OUTDOORS: 2 weeks after the last spring frost
SUN NEEDS: 6 to 8 hours per day
WATER NEEDS: high
ROTATION: follow a nitrogen-fixing crop
PLANTS PER SQUARE FOOT: 2

OVERVIEW: Corn takes up a lot of space in the garden, but many gardeners find it's worth the sacrifice. Fresh corn is sweeter and juicier than what you can buy at the store—in fact, it's so tasty it can be eaten raw, right off the stalk.

START: Corn is very sensitive to frost. It should be planted after all danger of frost has passed and when soil temperatures are at least 60°F. Corn does not transplant well, so it's best to sow seeds directly in the garden. Plant seeds 1½ to 2 inches deep and 6 to 12 inches apart. Because corn is pollinated by the wind, it should be planted in blocks of at least four rows rather than in long rows. For a continuous harvest, sow seeds in succession every 2 weeks.

GROW in a sunny area in loose, well-draining soil. Corn is a heavy feeder, especially of nitrogen, so you'll need to fertilize the soil well before and while growing your plants. The autumn before sowing seeds, enrich the soil with a layer of compost or aged manure. Then, once you've planted your crops, fertilize the soil every 2 weeks with a complete organic fertilizer, such as fish emulsion. When plants are 8 inches tall, side-dress them with a nitrogen-heavy fertilizer; repeat when plants are 18 inches tall. Ensure your plants get plenty of water, too. Water-related stress during pollination often results in ears with missing kernels. Mulching the soil around plants will go a long way in helping to retain soil moisture. Growing beans along with your corn will also help fix nitrogen in the soil; see page 61 for information on the Three Sisters method of planting.

HARVEST: When the tassels on the top of the ears turn brown, peel back part of the husk and pinch a kernel; if a milky white liquid squirts out, the ears are at peak ripeness and ready to harvest. To harvest ears, pull them downward and twist them off their stalks. Ears on the same stalk usually ripen a few days apart. If the silk is completely dry or the sheath is yellow or faded green, the ear is past its prime. Corn begins to lose its sweetness soon after harvesting, so eat it the same day you harvest it, if possible.

CONTAINERS: Corn is not well-suited for growing in containers.

PROBLEMS: Corn is susceptible to several garden pests and diseases, including cutworms, cucumber beetles, deer, flea beetles, and smut.

cucumbers

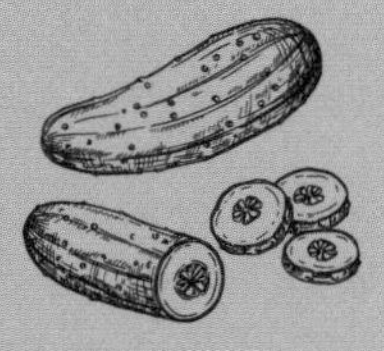

FAMILY: Cucurbitaceae
GROWING SEASON: warm season
PLANTING: start indoors or direct-sow outdoors
START INDOORS: 1 week before the last spring frost
SOW OUTDOORS OR TRANSPLANT: 2 weeks after the last frost
SUN NEEDS: 6 to 8 hours per day
WATER NEEDS: high
ROTATION: avoid following other crops in the gourd family
PLANTS PER SQUARE FOOT: 1 (bush), 2 (trellised)

OVERVIEW: A tropical vegetable, cucumbers grow best in warm weather. Plants are so sensitive to frost that they shouldn't be set outside until soil temperatures are consistently in the 70°F range. Cucumbers grow in two different forms: vining and bush. Vining cucumbers scramble along the ground or climb support structures such as trellises or fences, while bush types grow as compact plants. Of the two, vining cucumbers tend to yield more fruit. Bush varieties are perfectly suited to containers and small gardens.

START: Cucumbers should be direct-sown or transplanted into the garden no earlier than 2 weeks after the last spring frost. Plant seeds 1 inch deep every 3 inches, then thin them to 6 to 9 inches apart for vining types and 12 inches apart for bush types. Cucumbers do not like to be transplanted, but if you want to get a head start, you can try starting seedlings indoors 3 weeks before it's time to move them to the garden. If you start seeds indoors, be sure to provide them with bottom heat by using a heating mat or setting the seeds on top of a warm refrigerator.

GROW: If you think about it, cucumbers are mostly water, which explains why they need so much of it to grow! If you remember one thing about growing cucumbers, let it be to water them consistently—about 1 inch per week—and never let the soil dry out. When watering your plants, prevent leaf diseases by aiming for the soil and not the foliage. Mulch the soil to retain moisture and side-dress plants with compost or aged manure to provide them with enough nutrients. One week after the flowers bloom, apply a 5-10-10 liquid fertilizer to the soil and reapply it again every 3 weeks.

HARVEST cucumbers when they are green, firm, and crisp. Using a knife or clippers, cut the stem just above the fruit. Avoid pulling them by hand, which can damage the plant.
Try not to leave cucumbers on the vine too long to prevent fruits from developing tough skin and a bitter taste. Pickling cucumbers are best harvested when they are small; slicing cucumbers should be harvested when they are 6 to 8 inches long.

CONTAINERS: Bush cucumbers grow well in containers that are at least 10 inches deep.

PROBLEMS: Common problems with cucumbers include aphids, cucumber beetles, flea beetles, mildew, squash vine borers, wilt, and whiteflies. If fruits aren't developing because flowers aren't being pollinated naturally, try pollinating the flowers by hand (see page 137).

dill

FAMILY: Apiaceae
GROWING SEASON: cool season
PLANTING: direct-sow outdoors
SOW OUTDOORS: after the last spring frost
SUN NEEDS: 6 to 8 hours per day
WATER NEEDS: high
ROTATION: follow beetroot
PLANTS PER SQUARE FOOT: 9

OVERVIEW: Dill is a cold-hardy, tall-growing plant with feathery leaves and yellow umbrella-shaped flowers. Its feathery green leaves are enjoyed as an herb often called dill weed, while its dried seeds are used as a spice. Dill weed is used to flavor many dishes, including salads, veggies, meats, and sauces. Dill seed is used to flavor pickles, sauerkraut, and coleslaw. Dill is often used to attract beneficial insects to the garden and is a host plant for the black swallowtail butterfly caterpillar.

START: Dill should be sown directly in the garden. Not only does it not transplant well, but it also grows quickly, so there is no need to start it early indoors. Dill is very easy to grow—just moisten the soil and broadcast the seeds. Press the seeds firmly into the ground but avoid covering them with soil, as they need light to germinate. When the seedlings are 3 inches high, thin them to 6 to 12 inches apart. To prevent weeds and conserve moisture, add a layer of mulch to the soil surrounding the plants. For a steady supply of dill, sow seeds every 2 to 3 weeks until midsummer. Dill will reseed itself readily, so plant it where you can allow it to grow for several years.

GROW in full sun and fertile, moist, well-draining soil. Water consistently throughout the growing season to keep plants from drying out.

HARVEST fresh leaves as needed during the growing season once plants are at least 8 inches tall. Leaves have the best flavor just before the flowers open, about 65 to 70 days after sowing. Dill seeds are ready to harvest about 90 days after sowing. When the seeds turn brown, cut off the flower heads and place them in a paper bag. Gently shake the bag and collect the seeds that fall to the bottom of the bag.

CONTAINERS: Dill will grow easily in a container that is at least 12 inches deep.

PROBLEMS: Dill is susceptible to parsley caterpillars and tomato hornworms. Either handpick them off or spray plants with *Bacillus thuringiensis* (Bt).

garlic

FAMILY: Amaryllidaceae
GROWING SEASON: cool season
PLANTING: direct-sow outdoors
SOW OUTDOORS: 2 weeks before the first autumn frost
SUN NEEDS: 6 to 8 hours per day
WATER NEEDS: low
ROTATION: avoid following onion or cabbage family crops
PLANTS PER SQUARE FOOT: 4 (large varieties); 9 (small varieties)

OVERVIEW: Garlic is one of the easiest crops to grow. The timing for planting it varies widely depending on where you live; in most places, it is planted in the autumn and harvested the following summer. When sourcing garlic to plant, buy bulbs from a seed company or a local nursery; the garlic sold in grocery stores is often treated to prevent sprouting. There are two types of garlic—soft neck and hard neck. Hard neck varieties are usually hardier than soft neck varieties and are the best option for northern gardeners.

START: To plant, break the bulbs apart into individual cloves, keeping the paper husk intact. Choose the largest, healthiest-looking cloves and plant them 2 inches deep, 4 inches apart, with the blunt end facing down and the pointed end facing up. Cover with 1 to 2 inches of compost and several inches of mulch to keep them from freezing or coming out of the ground.

GROW garlic in a sunny location in rich, well-draining soil. Once the leaves emerge in the spring, remove the thick layer of mulch and side-dress the plants with an organic fertilizer high in nitrogen. Water the plants every 3 to 5 days while the garlic is growing, then taper off a few weeks before harvesting the bulbs.

HARVEST: In early to mid-June, garlic will send up a scape (a curly, flowering stalk) from its center. If left unharvested, the scape will form a flower and will eventually go to seed. The scape should be harvested to help the plant put its energy toward bulb development rather than seed production. To harvest the scape, wait until the center stalk grows above the rest of the plant and begins to curl or spiral. Then cut the stalk as far down as possible without damaging the surrounding leaves. Garlic bulbs should be harvested when the lower leaves turn yellow (in soft neck varieties) or fall over (in hard neck varieties). The only way to be sure your garlic is ready is to dig up a few bulbs to check their progress. When the cloves fill out their skins, it's time to harvest them. Harvesting too soon results in smaller cloves that don't store well.

CURE: After harvesting, you can eat garlic fresh or cure it for storage. To cure garlic, start by brushing off any dirt. To prevent rot, avoid washing the bulbs or getting them wet. Bundle three or four bulbs together, tie them with twine, and hang them upside down (with the bulbs on top and the leaves dangling below) in a cool, dry place with good air circulation for 3 to 4 weeks. Once the tops and roots have dried, snip them off, and clean the garlic by removing the outermost layer of papery skin, being careful not to expose the cloves.

CONTAINERS: Garlic grows well in containers at least 8 inches deep.

PROBLEMS: Common problems include rust and rot.

green beans

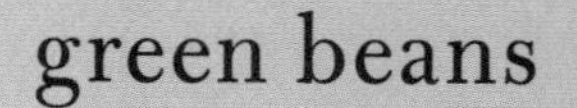

FAMILY: Fabacaeae
GROWING SEASON: warm season
PLANTING: direct-sow outdoors
SOW OUTDOORS: 2 weeks after the last spring frost
SUN NEEDS: 6 to 8 hours per day
WATER NEEDS: moderate
ROTATION: avoid following other legumes or planting before or after onion family crops
PLANTS PER SQUARE FOOT: 4

OVERVIEW: Green beans are one of the quickest and easiest crops to grow. Though most are green, they come in a variety of other colors, such as gold, purple, red, and yellow. Green beans can be classified as bush beans or pole beans. Bush beans grow into compact plants that require little to no support; pole beans grow as long vines that reach 10 to 15 feet in length and need support from trellises. Bush beans require less maintenance but also yield a smaller harvest than pole beans.

START: Beans grow best when sown directly in the garden. Bean seeds are sensitive to frost and like air temperatures between 70°F and 80°F and soil temperature above 60°F. Plant bush bean seeds 1 inch deep and 3 to 6 inches apart, thinning to 6 inches apart. For a continuous crop, sow every 2 weeks until 2 months before the first autumn frost. Plant pole bean seeds 2 inches deep and 10 inches apart. Before sowing seeds, be sure to set up appropriate support structures, if needed. Bush beans can reach 12 to 24 inches in height, whereas pole beans can grow up to 15 feet tall.

GROW in full sun in well-draining soil with a pH between 6 and 7. Add compost or aged manure to beds the autumn before planting. Mulch to help soil retain moisture and keep roots cool. Water well and consistently, up to 2 inches per week. To increase yields, side-dress plants with more compost and aged manure halfway through the growing season. Pinch pole beans when they reach the top of supports to encourage them to produce more beans instead of growing taller.

HARVEST beans when they are young and tender, before the seeds start to swell. If seeds are large or swelling inside the pods, there's a good chance the beans are past their prime. Snap or cut beans off the plants daily to encourage higher yields and pick in the mornings when sugar levels are high. The rule with beans is, the more you pick, the more you get, so don't hold back!

CONTAINERS: Pole beans will grow in containers that are at least 10 inches deep, and bush beans will grow in containers that are 6 to 7 inches deep. The easiest way to support pole beans growing in containers is to insert a stick, pole, or tomato cage into the pot.

PROBLEMS: Green beans can be affected by a number of pests and diseases, many of which remain in the soil for several years. The best way to prevent them from spreading or recurring is to avoid working with your beans when the foliage is wet and to practice crop rotation. If beetles or bugs are a problem, pick them off and toss them into a jar of soapy water.

kale

FAMILY: Brassicaceae
GROWING SEASON: cool season
PLANTING: start indoors or direct-sow outdoors
START INDOORS: 4 to 6 weeks before the last spring frost
SOW OUTDOORS OR TRANSPLANT: 2 to 4 weeks before the last frost
SUN NEEDS: 3 to 4 hours per day
WATER NEEDS: moderate during growing season; low after frost
ROTATION: avoid following cabbage family crops
PLANTS PER SQUARE FOOT: 1

OVERVIEW: Kale is an easy-to-grow cool-season member of the cabbage family. It grows best in spring and autumn, when temperatures are between 60°F and 65°F. Kale is extremely hardy—not only can it survive the early months of winter without protection, but it also tastes better after exposure to light frost. Kale will continue to grow until temperatures go below 20°F.

START: Kale is easy to grow from seed, either directly sown in the garden or started indoors and transplanted. You can also buy starts from a nursery. Direct-sow seeds ¼ to ½ inch deep, 12 inches apart. Keep the soil evenly moist until the seeds germinate, and thin the weakest seedlings a few weeks later. For an autumn crop, start seeds indoors 12 to 14 weeks before the first autumn frost, and transplant seedlings 6 to 8 weeks later. When transplanting seedlings, space them 12 inches apart.

GROW kale in full sun or partial shade and in loamy, well-draining soil. Before planting, enrich the soil with compost or aged manure to feed the soil and help it retain moisture. Kale likes moist but not soggy soil; mulch to keep roots cool and moist. When plants are about 5 inches tall, side-dress them with an organic fertilizer high in nitrogen.

HARVEST: Kale is a cut-and-come-again crop that can be harvested as soon as the leaves are 5 to 6 inches long. Harvest from the outside and avoid removing the terminal bud at the top center of the plant. Pick no more than one-third of the plant at a time, leaving enough foliage to allow the plant to continue to grow. To avoid damaging the plant, cut stems with scissors or a sharp knife.

CONTAINERS: Kale grows well in containers at least 10 inches deep.

PROBLEMS: Aphids, black root, cabbage caterpillars, and flea beetles are common pests and diseases. Heat is also problematic, as it can toughen leaves and cause plants to bolt.

leeks

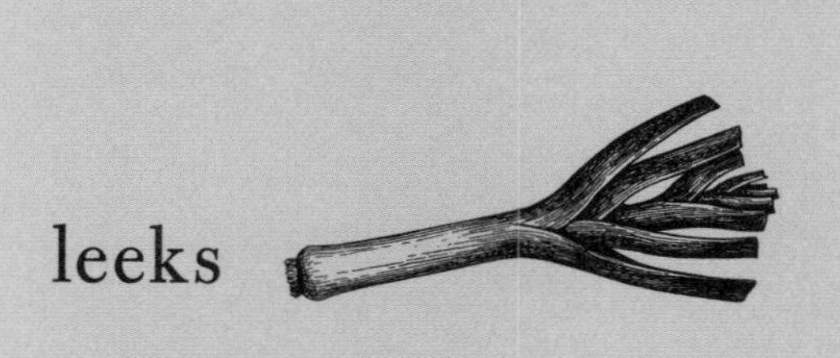

FAMILY: Alliaceae
GROWING SEASON: cool season
PLANTING: start indoors or direct-sow outdoors
START INDOORS: 10 to 12 weeks before the last spring frost
SOW OUTDOORS OR TRANSPLANT: after the last spring frost
SUN NEEDS: 6 to 8 hours per day
WATER NEEDS: moderate
ROTATION: avoid following onion family crops
PLANTS PER SQUARE FOOT: 4 (large varieties), 9 (small varieties)

OVERVIEW: Leeks are a hardy, cold-season veggie that can survive frosts. A biennial member of the allium tribe, leeks are closely related to garlic, onions, shallots, and chives. Their white shanks have a mild, sweet onion flavor. Although leeks prefer cool, rainy climates, they grow fine in warmer, drier climates when earthed up with soil and given plenty of water.

START: Because leeks have a long growing season of about 120 to 150 days, some gardeners like to start them from seed indoors. To start them indoors, sow seeds about 1 inch apart in trays or pots, covering them with a dusting of soil. To encourage seedlings to grow strong, healthy roots, use scissors to give them a weekly haircut. Trimming their leaves to a height of 2 inches above soil level will force them to direct their energy toward the roots. When temperatures are at least 45°F, transplant seedlings into the garden.

To produce leeks with long white shafts, they must be blanched, or covered and hidden from the sun. The best way to do this is by planting seedlings 4 inches apart in a narrow trench, 6 to 8 inches deep, and tucking them into the trench up to the base of their first green leaves. As the leeks grow, earth them up by mounding soil around the base of each plant every 2 weeks.

GROW: Leeks grow best in fertile soil that has been enriched with organic matter. Before planting, add plenty of compost to the garden bed. Water evenly and consistently, up to 1 inch per week. A layer of mulch will help conserve soil moisture and deter weeds.

HARVEST: A leek is ready to harvest when the base of the stalk is 1 to 2 inches in diameter. A good-quality leek will have a firm, white shaft more than 3 inches long. Harvest leeks by slipping a trowel or fork underneath the plant while pulling up on the leaves. Shake off the soil, then trim the roots and all but 2 inches of the leaves.

CONTAINERS: Leeks grow well in containers that are at least 18 inches deep. Fill them only about two-thirds full of soil when you're getting started so you can earth up plants as they grow.

PROBLEMS: Common pests and diseases include onion maggots, thrips, and white rot.

lettuce

FAMILY: Asteraceae
GROWING SEASON: cool season
PLANTING: start indoors or direct-sow outdoors
START INDOORS: 6 weeks before the last spring frost
SOW OUTDOORS OR TRANSPLANT: 2 weeks before the last spring frost
SUN NEEDS: 3 to 4 hours per day
WATER NEEDS: low to moderate
ROTATION: avoid planting after endive, escarole, and artichokes
PLANTS PER SQUARE FOOT: 16 (leaf), 4 (head)

OVERVIEW: Lettuce is one of the easiest and quickest vegetables to grow. As a cool-season crop, it grows best in spring and autumn when temperatures are between 45°F and 65°F. Lettuce comes in many different shades and varieties and can be harvested as microgreens, as baby greens, half grown, or full grown.

START: Whether starting seeds indoors or outdoors, sow them the same way. Plant seeds ¼ inch deep. Cover them lightly with soil so light can penetrate the surface and help the seeds germinate. Thin plants when they have three or four true leaves. If transplanting seedlings you started yourself, wait until each plant has four to six leaves and a well-developed root system. For a continuous harvest, sow seeds every 2 weeks until temperatures exceed 80°F, then start again 8 weeks before the first autumn frost.

GROW lettuce in a sunny spot with loose, well-draining soil. Lettuce seeds are tiny, so be sure to remove rocks, stones, and clods of dirt that may inhibit germination. During summer, grow lettuce beneath taller plants, like corn or tomatoes, to provide shelter and prevent bolting. Enrich the soil with compost or aged manure 1 week before setting out seeds or transplants and feed it a few weeks later with a high-nitrogen fertilizer such as alfalfa meal. Add a layer of mulch to help retain moisture, suppress weeds, and moderate soil temperatures. Keep soil moist but make sure it's also draining well. Lettuce does not need a lot of water, but you also don't want it to wilt.

HARVEST in the morning when the leaves are crisp. Some varieties of lettuce are grown to be eaten as baby greens, which can be harvested when the leaves are 2 to 4 inches long. To harvest baby greens, use scissors to cut the leaves at soil level. Most lettuces can be harvested as a cut-and-come-again crop by removing exterior leaves and allowing plants to continue to grow from their center stems. To harvest an entire head of lettuce, cut it at the base with a sharp knife.

CONTAINERS: Lettuce grows well in containers at least 4 to 6 inches deep.

PROBLEMS: Slugs and aphids are two of the most common problems with growing lettuce. Heat is also a concern, as it causes lettuce to bolt quickly.

melons

FAMILY: Curcubitaceae
GROWING SEASON: warm season
PLANTING: start indoors, direct-sow outdoors, or purchase transplants
START INDOORS: 3 weeks before the last spring frost
SOW OUTDOORS OR TRANSPLANT: 2 weeks after the last spring frost
SUN NEEDS: 6 to 8 hours per day
WATER NEEDS: moderate
ROTATION: avoid following cucumbers, pumpkins, and summer or winter squash
PLANTS PER SQUARE FOOT: 1 per square foot (trellised); 1 per 4 square feet (not trellised)

OVERVIEW: Growing up in the South, I thought watermelons were one of the easiest crops to grow. Now that I live in a northern climate, I understand just how finicky they can be. Melons of all kinds—cantaloupe, honeydew, and watermelon—need long, warm summers to grow well. Although they can grow in northern regions, they require extra care and attention to produce healthy fruit.

START: In northern climates, get a head start on the growing season by starting melons indoors or buying transplants from a local nursery. To start plants indoors, sow two seeds in individual 3- to 4-inch pots and place them on a heat mat to keep the soil at 75°F. Transplant melons, spacing them 2 feet apart, after the last spring frost, when daytime temperatures are consistently above 70°F. If you live in a warm climate, sow seeds directly outdoors ½ inch deep, 2 feet apart.

GROW melons in full sun and moist, fertile, well-draining soil. Melons are particular about soil pH, preferring soil in the 6.5 to 7.0 range. They also need soil that drains well but retains moisture, so be sure to work plenty of compost into the soil before planting. In northern climates, use row covers to keep plants warm and to deter pests, being sure to remove covers as soon as plants start to blossom. Provide melons with plenty of water, especially when they're blooming and when their fruits are starting to develop. Decrease watering when it's almost time to harvest, as too much water will dilute a melon's flavor. If space is limited, grow melons vertically, using homemade slings (see page 126) to keep the fruit from falling and cracking.

HARVEST: Melons are ready to harvest when small cracks appear on their stems. Once the stems shrivel, melons will break off with the slightest twist. If more than a little effort is needed, the melon is neither ripe nor ready to be picked.

CONTAINERS: Melons grow well in large containers that are at least 12 inches deep.

PROBLEMS: Melons can be hard to grow because of their temperature needs. They also suffer from pests and diseases, such as aphids, cucumber beetles, and powdery mildew.

mint

FAMILY: Lamiaceae
GROWING SEASON: perennial
PLANTING: purchase transplants
TRANSPLANT: after the last spring frost
SUN NEEDS: 4 to 6 hours per day
WATER NEEDS: high
ROTATION: n/a
ZONES: 3–8
PLANTS PER SQUARE FOOT: 1

OVERVIEW: Mint is one of the easiest and most popular perennial herbs to grow. It's also the most difficult to contain. While a tiny seedling might start out looking well behaved and manageable, it will sprawl like the dickens if you let it. It doesn't take long for gardeners to figure out that mint needs boundaries, but once you contain it, it makes a wonderful addition to the garden. Mint can be used in a variety of ways, from cooking to brewing tea, making medicine, mixing mojitos, and attracting pollinators. Mint also comes in a variety of flavors, including spearmint, peppermint, chocolate mint, and apple mint.

START: Mint is not reliable when grown from seed. Because different varieties cross-pollinate, offspring are rarely true to type. The best option is to buy transplants from a local nursery so you know exactly what to expect. Plant seedlings 12 inches apart and either sink a barrier around them or grow them in containers.

GROW: Mint is not all that fussy to grow. It thrives in well-draining fertile soil and in light shade, where its roots can stay moist without becoming waterlogged.

HARVEST: Mint can be harvested at any time once the plant is 3 to 4 inches tall. Simply pinch off stems as needed. For a large harvest, wait until just before the plant blooms, when the flavor is most intense. Mint can be enjoyed fresh or dried for tea.

CONTAINERS: Mint grows well in containers that are at least 8 inches deep. Container-grown plants may need protection in the winter to ensure they come back the next year.

PROBLEMS: As far as pests and diseases go, rust and wilt are the two biggest concerns. As a fast grower, it may outgrow its pot quickly. If it stops growing or becomes woody, transplant it to a bigger container.

onions

FAMILY: Alliaceae
GROWING SEASON: cool season
PLANTING: direct-sow onion sets outdoors
SOW OUTDOORS: 1 to 2 weeks before the last spring frost
SUN NEEDS: 6 to 8 hours per day
WATER NEEDS: moderate
ROTATION: avoid following onion family crops
PLANTS PER SQUARE FOOT: 9

OVERVIEW: Onions are a hardy cold-season crop that grows best in autumn or spring. They're slow-growing and take several months to mature, but they're worth the effort. Not only are they practical and useful for cooking, but they also keep for up to 3 months when stored properly.

START: You can start onions from seeds, seedlings, or sets, although beginner gardeners typically grow them from seedlings or sets. Sets are small bulbs that produce mature onions in about 14 weeks. Not only do they have a higher success rate than seeds and seedlings, but they're also resistant to frost. If starting with sets, sow them 1 inch deep and 3 inches apart, with the pointed side level with the soil surface. Choose smaller sets that are approximately ¾ inch in diameter, as larger ones tend to develop stiff necks and bolt too quickly.

GROW onions in fertile, loose soil that is rich in organic matter. Onions like full sun, good drainage, and slightly acidic soils with a pH between 6.2 and 6.8. As heavy feeders, they need a lot of nourishment to grow into big, healthy bulbs. Before planting sets, add compost or aged manure to the soil, then feed it with a nitrogen-rich fertilizer every few weeks until the onions emerge from the surface. Cover the soil with a light layer of mulch and water about 1 inch per week.

HARVEST: Onions are ready to harvest when their tops turn yellowish brown and fall over. To harvest them, wait for a dry day, as wet onions don't cure well, then dig them up carefully with a trowel.

CURE: If you plan to store your onions, you'll need to cure them first. Spread them out on a clean, dry surface in a well-ventilated location for 2 to 3 weeks, or until the top necks are completely dry and the outer skins are crisp. Then cut the tops off within one to two inches of the bulb and store the onions in a wire basket or wooden crate in a cool, dark, dry place.

CONTAINERS: Onions can be grown in pots that are at least 10 inches deep.

PROBLEMS: Common pests and diseases afflicting onions include onion maggots, thrips, and white rot.

parsley

FAMILY: Apiaceae
GROWING SEASON: cool season
PLANTING: start indoors or direct-sow outdoors
START INDOORS: 10 to 12 weeks before the last spring frost
SOW OUTDOORS OR TRANSPLANT: 3 to 4 weeks before the last spring frost
SUN NEEDS: 3 to 4 hours per day
WATER NEEDS: low
ROTATION: avoid following carrots, celery, or parsnips
PLANTS PER SQUARE FOOT: 2

OVERVIEW: Parsley is a biennial herb that is typically grown as an annual. In addition to being one of the most popular culinary herbs, it's also one of the healthiest, containing three times as much vitamin C by weight as oranges! The two most common varieties are curly and flat-leaf (Italian). Curly parsley is typically used as a garnish, while flat-leaf parsley is used in cooking, thanks to its bold, aromatic flavor.

START: Parsley germinates slowly and sporadically. For best results, soak the seeds in water overnight before planting. To get a head start on the growing season, start seeds indoors 10 to 12 weeks before the last spring frost. Otherwise, direct-sow them outdoors 3 to 4 weeks before the last spring frost. Plant seeds ¼ inch deep and 6 inches apart. Because seeds take so long to germinate (up to 3 weeks), many gardeners prefer starting with transplants.

GROW: Parsley can grow in full sun to partial shade and prefers rich, moist, well-draining soil. Before planting, ensure your soil is packed with nutrients by adding several inches of aged manure or rich organic compost. Parsley makes a great companion plant to several veggies such as asparagus, carrots, corn, onions, peas, peppers, and tomatoes. Not only does it enhance their flavors, but it also acts as a natural pest control.

HARVEST as needed, picking the outer leaves first so the inner ones can mature. Take a sprig or two at a time and cut them off at soil level. Cut parsley will keep longer if the stalks are kept in a small jar of water, like a flower bouquet.

CONTAINERS: Parsley grows well in containers that are at least 8 inches deep.

PROBLEMS: Crown rot, root rot, leaf spot, and botrytis blight are the most common problems for parsley and result from persistently wet soil. If infected, remove damaged plants, thin plants to improve air circulation, and refrain from overhead watering.

peas

FAMILY: Fabaceae
GROWING SEASON: cool season
PLANTING: direct-sow outdoors
SOW OUTDOORS: 4 to 6 weeks before the last spring frost
SUN NEEDS: 3 to 4 hours per day
WATER NEEDS: moderate during growing season; low after frost
ROTATION: avoid following other legumes or onion family crops
PLANTS PER SQUARE FOOT: 8

OVERVIEW: Peas are a low-maintenance cool-season crop that can be planted as early as the ground can be worked. Most gardeners grow one of three types of peas: sweet peas (with inedible pods and edible peas), snow peas (with edible pods and small, edible peas), and snap peas (with edible pods and large, edible peas).

START: Peas do not transplant well, so it's best to direct-sow them outdoors. Start seeds in early spring for a summer harvest and in late summer for an autumn harvest. Sow seeds 2 inches apart and thin to 4 to 6 inches apart when seedlings are 1 to 2 inches high. Peas can withstand cold temperatures, even snow, as long as temperatures don't dip into the teens. If it gets that cold, you may need to plant them again. Set up the appropriate type of support at the time of planting.

GROW peas in full sun, in evenly moist, well-draining soil. Add compost or aged manure to beds the autumn before planting. Mulch the soil as the weather gets warm to prevent plants from drying out. Since peas fix nitrogen in the soil, you shouldn't need to use a fertilizer, but if you do, be sure to avoid ones high in nitrogen. Otherwise, plants may have lush foliage but little to no flowering or fruit. Keep vines trained up a vertical support and avoid watering the foliage.

HARVEST peas when the pods are full and bright green. Once the pods start to lose their nice pea-green color and turn dull and brown, they've matured past their prime. Harvest pods carefully by picking or cutting them off their stems.

CONTAINERS: Peas grow well in containers 8 to 12 inches deep. Be sure to set up a trellis for them to climb. Bamboo stakes tied together with twine in a teepee shape work well.

PROBLEMS: Timing is important with peas, as they will stop flowering when air temperatures reach the high 80s. It's also important to prevent the plants from drying out; dehydrated plants will not produce flowers or pods.

peppers

FAMILY: Solanaceae
GROWING SEASON: warm season
PLANTING: start indoors or purchase transplants
START INDOORS: 8 to 10 weeks before the last spring frost
TRANSPLANT: 2 weeks after the last spring frost
SUN NEEDS: 6 to 8 hours per day
WATER NEEDS: high
ROTATION: avoid following tomatoes
PLANTS PER SQUARE FOOT: 1

OVERVIEW: Peppers are a classic warm season crop grown as perennials in tropical areas and annuals in temperate climates. Peppers come in many varieties, from sweet to bell to hot. Habanero, jalapeno, and Thai chilli are good choices if you want to add bold flavors to your favorite dishes. Peppers require a fairly long growing season, often up to 100 days, so the shorter your summer, the sooner you should start seeds indoors.

START: Peppers need heat to germinate and can be challenging to start from seed outdoors. Unless you live in a warm region with a long growing season, your best bet is to start seeds indoors or buy transplants from a nursery. Young peppers are sensitive to transplant shock, which you can avoid by transplanting them after all danger of frost has passed and when soil temperatures are at least 65°F. Gardeners in cold climates can warm their soil faster by covering it with a black plastic sheet for a few weeks before planting. When the time is right, space seedlings 12 inches apart and stake them at the time of planting.

GROW peppers in sandy, loamy soil that drains well. Before planting, enrich the soil with compost or aged manure, then mulch beds or containers to help conserve soil moisture and deter weeds. Weed and water regularly, being sure to keep the soil evenly moist. If you live in a warm climate, plant peppers so they'll get some shade on hot days; although they love heat, peppers are susceptible to sunburn! If blossoms appear early in the season, pinch them off—it may seem unnatural, but doing so will help the plants direct their energy into growing more fruit later in the season.

HARVEST: Most peppers change colors as they ripen and mature. When the fruit changes to its intended color, cut it from the stem with a sharp knife or a pair of scissors. Alternatively, pick them while they're still young and green—they won't taste as sweet, but harvesting them early will encourage your plants to produce more fruit.

CONTAINERS: Peppers are well-suited for containers that are at least 10 inches deep.

PROBLEMS: Peppers resist most pests but are susceptible to fungal diseases such as bacterial leaf spot, mosaic virus, southern blight, powdery mildew, blossom-end rot, and sun scald.

potatoes

FAMILY: Solanacaceae
GROWING SEASON: cool season
PLANTING: direct-sow outdoors
SOW OUTDOORS: 4 to 6 weeks before last spring frost
SUN NEEDS: 4 to 6 hours per day
WATER NEEDS: moderate
ROTATION: avoid following tomato family crops
PLANTS PER SQUARE FOOT: 2

OVERVIEW: The key to growing potatoes is making sure their tubers don't get exposed to sunlight. Otherwise, they turn green and produce a toxic substance that results in bitter, unpalatable potatoes that can make you sick if you eat too many. To prevent this, gardeners "earth up" their potatoes by piling soil over the base of their stems (where the tubers grow).

START: The best way to grow potatoes is to plant seed potatoes in the garden as soon as you can work the soil. Seed potatoes are not actually seeds—they are potato tubers saved from the previous year's harvest and stored over the winter. Purchasing certified seed potatoes from a garden store is the best way to ensure a healthy crop. You can also save some of your own harvest for replanting the next year, but don't use grocery store potatoes, as they're often treated with anti-sprouting chemicals. Seed potatoes come in a variety of different sizes. Smaller, golf-ball-sized tubers can be planted directly in the ground; larger tubers should be cut into 1½-inch pieces, each with at least two "eyes." To prepare seed potatoes, place them in a well-ventilated area for 1 to 2 days before planting them. When you're ready to plant, remove about 5 inches of soil, add 1 inch of compost or aged manure to the hole, place seed potatoes cut-side down, 6 inches apart, and cover them with soil.

GROW potatoes in cool, loose, well-draining soil. A few weeks after planting potatoes, add 4 inches of soil to create a hill above the ground, being sure to leave some of the plant exposed. Add a layer of mulch to help retain soil moisture, moderate soil temperature, and deter weeds. Continue this process every few weeks or as needed. Maintain even moisture, giving plants 1 to 2 inches of water per week. A few weeks before harvesting your potatoes, cut back on how much you water them, to help toughen them up and prepare them for storage.

HARVEST: Potatoes are ready to harvest when their foliage has completely died back. To test whether they are ready, dig one up and observe the skin. If the skin is tough and firmly attached, the potatoes are ready. If it is thin and rubs off easily, the potatoes aren't mature enough to harvest. When potatoes are ready, dig them up with a garden fork. If you plan to store your potatoes, cure them for 2 weeks by placing them in a cool, dark, dry place between 45°F and 60°F.

CONTAINERS: You can plant four to six seed potatoes in a large pot or other container that is at least 16 inches in diameter and 16 inches deep. (See page 191 for how to plant a sack of potatoes.)

PROBLEMS: Early blight is a common problem with potatoes, but it can be prevented by avoiding rotating them with tomatoes.

radishes

FAMILY: Brassicaceae
GROWING SEASON: cool season
PLANTING: direct-sow outdoors
SOW OUTDOORS: 4 to 6 weeks before the last spring frost
SUN NEEDS: 4 to 6 hours per day
WATER NEEDS: low
ROTATION: follow a legume family crop
PLANTS PER SQUARE FOOT: 16

OVERVIEW: Radishes are hardy cool-season veggies grown for their roots and leaves. They're one of the easiest crops to grow and do particularly well in the spring and autumn. Radishes are an ideal crop for small-space gardeners, as they require little space and can be grown between larger, taller plants. Fresh radish greens have a wonderful flavor and can be mixed in with other salad greens or used to make radish leaf pesto. They taste best when they're small and tender.

START: Before planting radishes, remove rocks, stones, and clods of dirt that may inhibit germination. Prepare the soil by adding aged manure or compost a few weeks before planting. As soon as the soil thaws and can be worked, sow seeds ½ inch deep and 1 inch apart. Water seeds well and keep them moist until they germinate, then thin to 3 inches apart. For a continuous harvest, sow seeds every 10 days until daytime temperatures are consistently 75°F or higher (high temperatures will cause radishes to bolt). In late summer you can begin sowing seeds again for an autumn harvest. Add a layer of mulch to the soil to keep the shoulders of maturing radishes covered and cool.

GROW radishes in cool, loose, moist soil that is rich in organic matter. Choose a sunny spot so radishes can direct their energy into growing roots rather than making more leaves. Provide radishes with even, consistent moisture to ensure healthy growth.

HARVEST: Small, short-season radishes are ready to harvest in 21 to 30 days, while larger, long-season varieties take 50 to 70 days to mature. Harvesting radishes when they're just big enough to eat is important, as the roots become woody and pithy when overripe. Most varieties of radishes should be harvested when their roots are approximately 1 inch in diameter at the soil line. Before storing them, cut off their tops, remove their root tails, and wash and dry them thoroughly.

CONTAINERS: Short-season radishes grow well in 6-inch-deep containers, while long-season radishes require containers 12 inches deep.

PROBLEMS: Flea beetles love radishes, although they don't seem to keep them from producing healthy roots. If the holes they create in leaves bother you, use floating row covers to protect crops, or sprinkle diatomaceous earth around the base of plants.

rocket

FAMILY: Brassicaceae
GROWING SEASON: cool season
PLANTING: direct-sow outdoors
SOW OUTDOORS: 1 to 2 weeks before the last spring frost or as soon as the soil can be worked, and again in late summer for an autumn or early-winter harvest
SUN NEEDS: 3 to 4 hours per day
WATER NEEDS: moderate and even
ROTATION: avoid following other crops in the brassica family
PLANTS PER SQUARE FOOT: 9

OVERVIEW: Also known as arugula, rucola, rucoli, and rugula, rocket is a fast-growing salad green with a spicy kick. There are a number of different types; in general, the thinner and spikier the leaves, the more peppery the flavor.

START seeds outdoors early or late in the growing season and harvest as soon as 4 weeks after sowing. Seedlings are cold hardy and will tolerate some frost. Plant ¼ inch deep and about 1 inch apart, or broadcast alone or with other salad greens and thin later. For a continuous harvest, succession-sow new seeds every 2 to 3 weeks.

GROW: Rocket isn't fussy and will grow in most soils. It has a shallow root system, though, and needs consistent watering to prevent it from drying out. Once you've planted the seeds, water regularly and try to keep the soil evenly moist. Thin seedlings to about 6 inches apart and toss the thinnings into salads. Rocket is a great plant to grow in the shade of taller plants, as it can quickly bolt in the heat of summer, which results in bitter-tasting leaves.

HARVEST rocket any time after the leaves are large enough to eat. Young, tender leaves taste sweet and mild, whereas older leaves are sharp and peppery in flavor. Once the seedlings are about 3 to 4 inches tall, you can either pull up the entire plant or selectively harvest the leaves as a cut-and-come-again crop. Alternatively, cut off the leaves just above the soil line and allow the plant to regrow. Rocket flowers are edible and taste great tossed on salads or sandwiches. Once the flowers start to bloom, however, the flavor of the leaves becomes bitter and intense.

CONTAINERS: Because of its small, compact root system, rocket grows well in shallow containers or flats, outdoors or indoors near a sunny window.

PROBLEMS: Rocket plants are favored by slugs, aphids, flea beetles, diamondback moths, and birds. Remove by hand any insect eggs you find and spray aphids off with a hose. Use beer traps to get rid of slugs. To prevent predation from flea beetles and birds, protect plants with row covers or netting, respectively.

rosemary

FAMILY: Lamiaceae
GROWING SEASON: perennial
PLANTING: use cuttings or purchase transplants
TRANSPLANT: after the last spring frost
SUN NEEDS: 6 to 8 hours per day
WATER NEEDS: low
ROTATION: avoid following cucumbers
ZONES: 7 and up
PLANTS PER SQUARE FOOT: 1

OVERVIEW: Rosemary is an evergreen shrub native to the Mediterranean. It grows best in warm, humid climates and can grow up to 5 feet tall in ideal conditions. Although it can grow as a perennial in Zones 7 and up, in colder areas it should be grown in pots that can be brought indoors for the winter.

START: Although rosemary can be grown from seed, germination rates are iffy, and seedlings grow extremely slowly. The easier and more predictable option is to purchase transplants from your local nursery or take cuttings from established plants. To grow plants from cuttings, cut stems that are about 2 inches long and remove the leaves from the bottom two-thirds of the cutting. Place the cuttings in a mixture of perlite and vermiculite, and water using a spray bottle until roots begin to grow. Once roots have developed, plant cuttings 1 to 3 feet apart.

GROW in full sun in well-draining, sandy soil with a pH between 6 and 7. Rosemary likes to be dry. Water thoroughly, but allow the soil to dry out between waterings. For container growing, consider using terra-cotta pots since they dry out faster than other options. If you bring rosemary indoors for the winter, be sure to keep it near a sunny window and away from cold drafts. It also helps to mist it daily with a spray bottle.

PRUNING: Rosemary typically won't need to be pruned unless it's overgrown, excessively woody, or you're trying to create a topiary shape or hedge. To prune it, clip off the faded flowers and snip off the top few inches of the stems, being careful not to cut too far into the old wood. If your rosemary is growing in a pot, it may become root-bound after a few years, which will cause it to produce less and less new growth. If this happens, repot the plant into a larger pot; otherwise, remove it from the container and carefully prune the roots back. Then place the plant in the same pot and add a fresh layer of potting soil.

HARVEST sprigs often once the plant is established. Use sharp pruning shears to harvest rosemary from mature plants with woody stems.

CONTAINERS: Rosemary grows well in containers at least 8 inches deep. Be sure to use a potting soil that drains well—one created for succulents would not be a bad idea.

PROBLEMS: Potted rosemary plants are prone to becoming root-bound and should be repotted once a year. Yellowing of the lower foliage is an indication that it's time to repot them. Whiteflies, spider mites, scale, mealybugs, and powdery mildew are common, particularly when drainage or air circulation are limited.

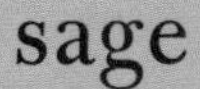

sage

FAMILY: Lamiaceae
GROWING SEASON: perennial
PLANTING: purchase transplants
TRANSPLANT: after the last spring frost
SUN NEEDS: 6 to 8 hours per day
WATER NEEDS: low
ROTATION: avoid following basil, cucumbers, or oregano
ZONES: 4–8
PLANTS PER SQUARE FOOT: 1

OVERVIEW: Sage is a lovely, low shrub with pale, velvet soft, grayish-green leaves. As a culinary herb, it is known for being highly aromatic, with a subtle, earthy flavor. Medicinally, it is used for all sorts of purposes, thanks to its antibacterial, antifungal, antiviral, and astringent properties. In Zones 4 through 8, sage is grown as a hardy perennial but is considered an annual in warmer climates due to its intolerance to heat and humidity. Sage makes a great companion plant to rosemary, cabbage, and carrots.

START: Although you can start sage from seeds, it can take a couple of years to fully mature. An easier option is to buy transplants from a nursery. When soil temperatures reach 65°F, plant seedlings 12 to 24 inches apart. Water young plants consistently and evenly. Once plants are well established, wait for the top 1 inch of the soil to completely dry out, then water it thoroughly. Overwatering sage makes it susceptible to mildew. Sage is fairly drought tolerant—if it begins to wilt, it will usually perk back up with a little water.

GROW sage in full sun in sandy, loamy, well-draining soil with a pH between 6.5 and 7. Before planting, add several inches of aged compost or rich organic matter to the soil. Once your sage is established, prune it back every year in early spring, cutting past its woody, thick stems.

HARVEST as needed by snipping the tips of the stems or by cutting them back to soil level. Enjoy sage fresh or dry it to use later for tea.

CONTAINERS: Sage grows well in containers that are at least 8 inches deep.

PROBLEMS: Sage is fairly resistant to pests and diseases, except in areas with high humidity, where it becomes susceptible to powdery and downy mildews.

spinach

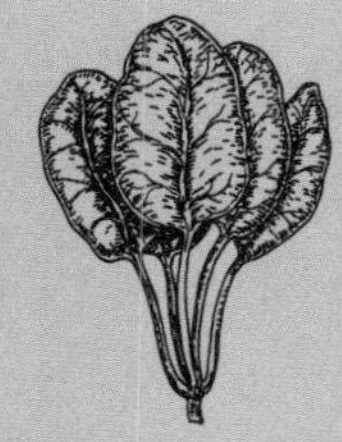

FAMILY: Amaranthceae
GROWING SEASON: cool season
PLANTING: direct-sow outdoors
SOW OUTDOORS: 4 to 6 weeks before the last spring frost
SUN NEEDS: 4 to 6 hours per day
WATER NEEDS: low but even
ROTATION: avoid following legumes
PLANTS PER SQUARE FOOT: 9

OVERVIEW: If your garden doesn't get much sun, spinach might be the ideal crop for you! A cool-season veggie, spinach can grow with as little as 4 to 6 hours of direct sunlight a day. It can also withstand light frosts. Just be sure to plant it in spring or autumn, as it tends to bolt in summer heat.

START: Although you can start seeds indoors, it's not ideal, as spinach can be difficult to transplant. A better option is to start seeds outdoors, sowing them ½ inch deep and 2 inches apart. Thin them to 4 inches apart when the plants are 2 inches tall. For a continuous harvest, sow new seeds every 2 weeks until air temperatures are consistently in the 70°F range. In early autumn, begin sowing seeds again for autumn and early-winter harvests.

GROW spinach in well-draining, sandy soil that is rich in organic matter. Before planting, work aged manure or compost into the soil. Water evenly and consistently and mulch around the base of plants to help retain soil moisture. When plants are 2 inches tall, side-dress them with a nitrogen-rich fertilizer to boost growth.

HARVEST spinach by cutting the entire plant off at the base or by picking the outer, older leaves first and gradually working your way to the center of the plant. Avoid waiting too long to harvest the leaves, as they become bitter soon after reaching maturity.

CONTAINERS: Spinach grows well in containers that are at least 8 inches deep.

PROBLEMS: Spinach bolts quickly when temperatures get too high. It is also prone to a few pests and diseases, such as leaf miners, downy mildew, and slugs.

strawberries

FAMILY: Rosaceae
GROWING SEASON: perennial
PLANTING: purchase transplants
TRANSPLANT: after the last spring frost
SUN NEEDS: 8 or more hours per day
WATER NEEDS: moderate
ROTATION: avoid following aubergines, beetroot, corn, peppers, and tomatoes
ZONES: 3–10
PLANTS PER SQUARE FOOT: 4

OVERVIEW: Success with strawberries requires understanding a little bit about their life cycle. Like many hardy perennials, they die back in the winter and return in the spring. After producing fruit, most varieties will produce runners, or daughter plants, which will send out their own runners. Although runners root themselves nearby, they stay connected to their mother plants. Generally speaking, a strawberry plant will bear more fruit if you clip back all but three of its runners. Although strawberries will come back year after year, they're most productive between years two and five; many gardeners replace them every few years as they become less productive.

START: Strawberries can be started from seeds or runners, but the surest way to grow them is by planting transplants from a local nursery. After the last spring frost and as soon as the soil can be worked, plant seedlings 6 inches apart in pots, raised beds, or the ground. If you would prefer to plant from seed, sow them directly in the garden after the last spring frost. It can take up to a month for strawberry seeds to germinate, so don't worry if it feels like nothing is happening after a few weeks. If you provide them with the right conditions, the seeds will sprout!

GROW strawberries in fertile, slightly acidic, well-draining soil. Because drainage is so important, gardeners often grow strawberries in raised beds or containers. Add a layer of compost or aged manure before planting and weed well, as weeds can easily out-compete the shallow roots of strawberries. Water plants sufficiently until they're well established, and add a layer of mulch to conserve moisture and deter weeds. Once your plants are covered in ripening berries, cover them with fine mesh netting to protect the fruit from birds.

HARVEST: To improve productivity, refrain from harvesting strawberries the first year. Instead, remove their blossoms to prevent the plants from fruiting. If they don't bear fruit, they'll direct their energy into developing strong roots, which will ensure higher yields in subsequent years. Harvest berries starting in year two when they are ripe, firm, and red.

CONTAINERS: Strawberries are ideally suited for raised beds and containers, including pots, grow bags, hanging baskets, and window boxes that are at least 6 inches deep.

PROBLEMS: Strawberries are prone to several pests and diseases, including aphids, beetles, birds, slugs, snails, botrytis, and powdery mildew.

summer squash

FAMILY: Cucubitaceae
GROWING SEASON: warm season
PLANTING: start indoors or direct-sow outdoors
START INDOORS: 2 to 4 weeks before the last spring frost
SOW OUTDOORS OR TRANSPLANT: 2 weeks after the last spring frost
SUN NEEDS: 6 to 8 hours per day
WATER NEEDS: high
ROTATION: avoid following with cucumbers, melons, pumpkins, or winter squash
PLANTS PER SQUARE FOOT: 2 per square foot (trellised); 1 per 2 square feet (not trellised)

OVERVIEW: Summer squashes are among the most prolific crops to grow. During the peak of the growing season, a single plant can produce several squashes a day. Summer squash varieties include courgettes, yellow squash, and crookneck squash.

START: Summer squash grows so quickly, there's really no need to start seeds indoors. But if you want to get a head start, sow them indoors 2 to 4 weeks before the last spring frost. Since squash seedlings don't transplant well, the best way to start them is in 4-inch biodegradable pots that can be transplanted along with the seedlings into the garden. To direct-sow seeds outdoors, wait until after the last frost date, when the soil temperature is 70°F, then plant them 2 feet apart. Water them thoroughly, and cover the soil with a layer of mulch.

GROW: Summer squash grows best in full sun and well-draining soil. Squashes are heavy feeders that thrive in fertile soil, so before planting, add aged manure or compost to the garden. After planting, make sure plants get plenty of water, especially when their fruits start to grow. Summer squash usually needs about 1 inch of water per week for maximum growth. Once their fruits appear, side-dress plants with a balanced organic fertilizer. If flowers are not being pollinated naturally, try hand-pollinating them using the technique described on page 137.

HARVEST: Most varieties of summer squash take about 60 days to mature and taste best when harvested young. To harvest, cut the squash from the vine with a sharp knife, leaving at least 1 inch of stem on the fruit. Harvest all fruit before the first autumn frost.

CONTAINERS: Summer squash grows well in large containers. Half-barrels are a great option, as are any containers that are at least 24 inches in diameter and 12 inches deep.

PROBLEMS: Common pests and diseases include squash bugs, squash vine borers, powdery mildew, and blossom-end rot.

sweet potatoes

FAMILY: Convolvulaceae
GROWING SEASON: warm season
PLANTING: direct-sow outdoors
SOW OUTDOORS: 2 to 4 weeks after the last spring frost
SUN NEEDS: 6 to 8 hours per day
WATER NEEDS: low to moderate
ROTATION: avoid following root crops
PLANTS PER SQUARE FOOT: 1

OVERVIEW: Sweet potatoes aren't grown from seed or starts but instead from "slips," which are the green vines that grow from the tubers when they sprout. You can buy slips from a local nursery or grow them from sweet potatoes.

START: Plant sweet potato slips when the soil temperature is at least 60°F and air temperatures are between 60°F and 85°F. To grow slips, purchase large, firm sweet potatoes in the autumn and store them in a cool, dark place until about 90 days before the last spring frost. Then place the sweet potatoes in containers filled with a few inches of potting soil. Cover each potato with a few more inches of potting soil and water them so the soil is damp but not soggy. Place on a heating mat or on top of a warm refrigerator to maintain a temperature between 75°F and 80°F. After 3 to 4 weeks, the potatoes will have grown slips with 6 to 12 inch long leaves and roots. A single sprouting sweet potato will grow at least 15 slips, which will create 15 plants and produce around 60 individual potatoes!

A few weeks before planting the slips outdoors, prepare the soil by mulching it with a layer of compost and warming it up with black plastic. When it's time to plant the slips, harden them off as you would any plant you've grown indoors. Then remove each slip from the potato by grasping it at its base and twisting. Plant the slips 12 to 18 inches apart, making sure the roots and ½ inch of the stem are covered with soil. Remove the lower leaves and water the slips generously.

GROW: Sweet potatoes prefer soil that is well drained and rich in organic matter. Once established, they'll tolerate dry soil, but they'll grow best if you keep them evenly moist with 1 inch of water per week. To keep tubers from splitting, stop watering plants about 3 to 4 weeks before harvesting them.

HARVEST: Sweet potatoes are ready between 100 and 140 days after planting, depending on the variety. Be sure to harvest them before the first autumn frost. Cut the vines and carefully dig up the tubers with a fork or by hand. Brush off excess dirt and either eat the potatoes immediately or cure them for storage. To cure sweet potatoes, store them in a warm, humid place for 14 days, then wrap them in newspaper and store them in a dark, cool spot for up to 6 months.

CONTAINERS: Small varieties can be grown in containers that are at least 12 inches deep.

PROBLEMS: Flea beetles, leafhoppers, and wireworms are the most common problems.

swiss chard

FAMILY: Amaranthaceae
GROWING SEASONS: cool season (will grow in warm season too)
PLANTING: direct-sow outdoors
SOW OUTDOORS: 2 to 3 weeks before the last spring frost
SUN NEEDS: 3 to 4 hours per day
WATER NEEDS: moderate
ROTATION: grow after legumes but not after beetroot or spinach
PLANTS PER SQUARE FOOT: 4

OVERVIEW: Swiss chard is a close relative of beetroot and is often grown as a summer substitute for spinach. Although typically grown as a cool-season crop in spring and autumn, Swiss chard can tolerate the warm temperatures that cause spinach to bolt. Swiss chard comes in a variety of colors and is valued as a superfood, rich in vitamins A, C, and K.

START: Before sowing seeds, soak them in water overnight to kick-start germination. Although you can start seeds indoors, most gardeners sow them directly outdoors, ½ inch deep and 2 to 4 inches apart. Like beetroot seeds, Swiss chard seeds are a cluster of seeds, meaning a few seedlings emerge when you plant one. When seedlings reach 3 inches tall, thin them to 6 inches apart by cutting the weakest plants at the base (pulling them can damage the roots of nearby seedlings). For an autumn harvest, plant seeds about 40 days before the first autumn frost.

GROW: Swiss chard grows best in well-draining soil enriched with aged manure or compost. Although it tolerates partial shade, it grows best in full sun. If the soil is nutrient-deficient, or if your plants seem to be growing slowly, apply a balanced organic fertilizer. Water plants evenly and consistently and use mulch to help retain soil moisture.

HARVEST: Swiss chard is best grown as a cut-and-come-again crop by harvesting a few outer leaves at a time. You can start harvesting leaves when they are 6 to 9 inches tall by cutting the leaves with a sharp knife, 1 inch above the ground. If you harvest often, Swiss chard will continue to produce new growth for months.

CONTAINERS: Swiss chard grows well in containers that are at least 8 inches deep.

PROBLEMS: The most common pests and diseases associated with Swiss chard include aphids, leaf miners, leaf spot, and slugs.

thyme

FAMILY: Lamiaceae
GROWING SEASON: perennial
PLANTING: use cuttings or purchase transplants
TRANSPLANT: after the last spring frost
SUN NEEDS: 3 to 4 hours per day
WATER NEEDS: low
ROTATION: avoid following cucumbers
ZONES: 4 and up
PLANTS PER SQUARE FOOT: 4

OVERVIEW: Thyme is a hardy perennial with dainty, fragrant leaves, thin woody stems, and adorable purple flowers that attract pollinators of all kinds. Like other Mediterranean herbs, thyme thrives on benign neglect and is fairly fuss-free. Thyme is also drought tolerant, and will grow in less fertile soils. It grows well in the ground, in beds, and in containers and can be left outdoors year-round in Zones 4 and up. Thyme makes a great companion to cabbage and tomatoes. It grows well in the same container as rosemary, as they share similar needs for sun, soil, and water.

START: Although you can grow thyme from seed, it's easier and faster to grow it from transplants or cuttings. Cuttings can be started indoors 6 to 10 weeks before the last frost date and transplanted into the garden when soil temperatures reach 70°F. To start a plant from a cutting, snip a 3-inch piece of stem from an established plant, apply rooting hormone to the cut end, and plant the stem in sterile sand or vermiculite. Within 6 weeks, after roots emerge, you can transfer the cutting to a pot or the garden. When planting in beds or the ground, cuttings should be spaced 12 inches apart.

GROW: Thyme thrives in full sun and well-draining, slightly sandy soil. Before planting, work some compost into the soil. Refrain from using fertilizers, as soil too rich in organic matter will produce plants that are large but less fragrant. Thyme requires regular watering the first year but grows best in dry soil thereafter. In cold regions, protect thyme during winter by applying a layer of mulch to its soil. To encourage vigorous growth and keep it from becoming too woody, prune plants by a third in the spring and again after flowering in the summer. After 3 to 4 years, divide or replace plants if they become woody and flavorless.

HARVEST: Snip sprigs of thyme as needed once plants are 6 to 8 inches tall. For drying, harvest just before the flowers bloom, when thyme's flavor is most intense.

CONTAINERS: Thyme grows well in containers that are about 6 inches deep.

PROBLEMS: Thyme is fairly resistant to pests and diseases. Poor drainage can cause botrytis rot or root rot. It is also sometimes attacked by aphids and spider mites.

tomatoes

FAMILY: Solanaceae
GROWING SEASON: warm season
PLANTING: start indoors or purchase transplants
START INDOORS: 6 to 8 weeks before the last spring frost
TRANSPLANT: after the last spring frost
SUN NEEDS: 6 to 8 hours per day
WATER NEEDS: moderate to high during growth; low as fruits mature
ROTATION: avoid following potatoes, peppers, and aubergines
PLANTS PER SQUARE FOOT: 1

OVERVIEW: Tomatoes come in all shapes and sizes and are classified as determinate or indeterminate. Determinate tomatoes, or bush tomatoes, grow 2 to 3 feet tall and their fruit ripens all at once. They don't require staking or caging and are ideal for growing in containers. Indeterminate tomatoes, or vining tomatoes, produce the largest types of slicing tomatoes and bear fruit all summer and until the first autumn frost. They require staking and are best suited to raised beds or in-ground gardens.

START seeds indoors, then harden off seedlings for a week before transplanting them into the garden. When transplanting seedlings, pinch off a few of the lower leaves, then place the root ball deep enough so the bottom leaves are just above the soil line. Roots will grow all along the plant's stem underground. Space seedlings 12 inches apart. Place tomato stakes or cages in the soil at the time of planting, and water plants well to reduce root shock. If you live in a warm climate with a long growing season, you can also sow seeds directly into the garden.

GROW in rich, well-draining soil. Water in the morning only and at soil level to prevent soilborne diseases. Mulch after transplanting to help retain moisture. When the tomato fruits are about 1 inch in diameter, side-dress the plants with compost, liquid seaweed, or fish emulsion every 2 weeks until harvest time. Avoid high-nitrogen fertilizers unless plants have yellow leaves; otherwise, plants will produce luxurious foliage but little to no fruit. If growing vining tomatoes, pinch off the suckers—the tiny new stems and leaves between the branches and main stem—to prevent plants from becoming overgrown at the expense of fruit production. As the plant grows, remove the leaves from the bottom 12 inches of the main stem.

HARVEST: Leave tomatoes on the vine as long as possible. Harvest tomatoes when they are firm and very red (or for tomatoes of other colors, when they turn the correct color). If the temperature starts to drop before your tomatoes ripen, harvest them and store them in a brown paper bag until they ripen. If you add a banana to the bag, they'll ripen faster. (Or cook the green tomatoes.)

CONTAINERS: Choose bush or dwarf varieties and grow them in a large pot that is at least 20 inches in diameter. Plant one tomato plant per pot.

PROBLEMS: Blights, blossom-end rot, flea beetles, Fusarium wilt, mosaic virus, nematodes, and Verticillium wilt all plague tomatoes.

winter squash

FAMILY: Curcubitaceae
GROWING SEASON: warm season
PLANTING: start indoors, direct-sow outdoors, or purchase transplants
START INDOORS: 4 weeks before last spring frost
SOW OUTDOORS OR TRANSPLANT: 3 to 4 weeks after the last spring frost
SUN NEEDS: 6 to 8 hours per day
WATER NEEDS: high
ROTATION: avoid following cucumbers, melons, or summer squash
PLANTS PER SQUARE FOOT: 1 (trellised); 1 per 2 square feet (not trellised)

OVERVIEW: Winter squashes are frost-sensitive, warm-season annuals. The trick to growing them is having enough warm weather to support their growth from seed to fruit. Not only do winter squashes not germinate well in cold soil, but they also need at least 3 months of frost-free time to grow. Winter squashes include acorn, butternut, delicata, hubbard, kabocha, spaghetti, and pumpkin, to name a few.

START: In warm regions, you can direct-sow seeds as soon as the soil is at least 70°F. In cooler regions, it's a good idea to start seeds indoors or buy transplants from a local nursery. If you decide to start seeds indoors, use biodegradable pots that can be transplanted with the seedlings in the garden. Also, use a heat mat to maintain a temperature of 70°F during germination and early growth. If direct-sowing seeds, plant two presoaked seeds in the center of a 2-square-foot planting area. Leave a 2-inch depression around the seeds to hold lots of water. If both seeds sprout, thin the weaker one.

GROW squash in loose, well-draining soil rich in organic matter. Before planting, prepare the site by adding compost or aged manure to the soil. Winter squash plants take up a lot of space, so if space is limited, train them up a 5- to 8-foot-tall trellis or fence. Winter squash requires a lot of water in hot weather and grows best when the soil is kept evenly moist. Side-dress squash with compost every couple of weeks during the growing season. If fruit development is slow or nonexistent, hand-pollinate the female flowers using the technique described on page 137.

HARVEST squashes when the skin feels hard, before the first autumn frost. If you can't indent the rind with your thumbnail and the vines are hard and dry, the squash is ripe and ready to be harvested. Use a sharp knife to cut the squash from the vine, leaving 2 to 3 inches of the stem on the squash.

CONTAINERS: Bush-type winter squash can be grown in containers that are at least 10 inches deep and near a trellis.

PROBLEMS: The most common pests and diseases associated with winter squash include cucumber beetles, squash borers, squash bugs, blossom-end rot, mosaic virus, and powdery mildew.

acknowledgments

So many wonderful people helped bring this book to life, I really can't say thank you enough.

To everyone at Mariner Books, for all their hard work and dedication. To my editor Stephanie Fletcher, for encouraging me to write this book during the height of the pandemic, when the world was locked down and so many were seeking solace in the garden. To Rebecca Springer, for contributing her attention to detail, laser-focused editing, seamless project management, and collaborative work style. To Ashley Lima and Melissa Lotfy, for turning my words and photos into *another* beautiful book!

To my literary agent, Julia Eagleton, for encouraging me to write a second book and for connecting me to the right publishers and editors in the U.S. and overseas.

To my family, for supporting me every step of the way. To my parents, for believing in me and always letting me chase my many dreams, ambitions, and ideas. To Eloise, for being my gardening buddy and helping me create the projects in the Playing section, and to Benjamin, for being my second pair of eyes, lending his visual and design prowess, and figuring out how to work a tripod and digital camera so he could help shoot the cover photo. To my husband Scott, for holding down the fort while I typed feverishly in the basement and for being the best in-house editor and wordsmith a girl could ask for.

To my grandparents (Mimi and Papa), for growing *me* in their garden, on homegrown corn, okra, melons, and tomatoes, and for showing me what it means to live simply, slowly, and sustainably. And to the farmers in West Africa, who taught me to grow food, nourish soil, and make compost with as few resources as possible—most of my confidence in working with my hands comes from the years I spent living with you!

And to the readers of *Simply Living Well*—this book wouldn't be possible without your continued support and encouragement. Thank you for giving me the opportunity to create and share and learn with you!

index

Page numbers in *italics* indicate illustrations.

D

E

F

G

V

W